IF THEY KNEW

IF THEY KNEW

A THRILLER

A.J. WOODING

Cover design by Christian Bentulan

Editing by Anette King

Editing by Heather Caryn

CONTENT WARNING

This novel contains themes and scenes that some readers may find distressing. For a complete list of content warnings, please visit: Ajwooding.com

For my son,
I love you more than anything in the whole world.
And for my husband,
Thank you for always believing in me, standing beside me, and supporting this dream from the very beginning. None of this would have been possible without you.
I love you.

Don't blame me if you never want to take a cruise again.

A.J. WOODING

PROLOGUE

MARCH 2013

With a loud blast, a ship's horn lets out a piercing *bwaah... bwaah*, and the low rumble in the water catches my attention for a brief second. This isn't the first time my mind has tricked me into thinking I was being rescued, so I sink back into myself. My eyes are almost entirely closed, and my skin feels like it's peeling off from the cold windburn and the dried saltwater. My body is so weak that I don't have the strength to lift my head and look around. Moments later, I feel icy cold water splashing on my legs, and the sound of a man yelling in my ear jolts me awake.

"Hello, can you hear me?"

I struggle to open my eyes; the sunlight is blinding. After my eyes adjust, I see a man in a jumpsuit and safety gear leaning over me. I'm shocked by the sight, and my dry throat burns as I gasp. My ears are ringing, and I feel like I'm underwater for a moment.

The man shouts again, "Can you hear me?"

Realizing this isn't a dream, I croak out, "Are you real?"

The man touches my arm and answers, "Sure am. My name is Brandon. I'm with the US Coast Guard. You're safe now. We've got a helicopter on the way to take you to a hospital."

I'm still in shock, so I just stare at him blankly. The man grabs my shoulder, and I flinch. He speaks softly, "Everything's going to be okay. Can you tell me your name?"

I'm still too stunned to speak.

He tries again, "Do you have someone we can call?"

This time, I managed to get the words out.

"No, I don't have anyone left. They're all dead."

One Year Later
March 2014

It's been a year since the "Parrish Cruise Line Disaster" gripped the world's attention. The media coverage of the horrific event spread like wildfire, growing so intense that its impact echoed around the world. The story of what happened on board was gruesome, terrifying, and almost unimaginable. The fallout was so severe that millions of people became afraid to take a commercial cruise. The PR nightmare was so devastating that the start-up cruise line couldn't recover and went bankrupt. Rumors about what happened on the ship circulated everywhere.

I'm currently sitting in a small diner, and I can hear a waitress standing just a few feet away gossiping about the disaster. She glances at me and smiles, and I can't help but think she has no idea what she is talking about. I'm not one to judge, but if I were, this lady looks like the type of woman who gossips just to stay relevant. She has box-dyed red hair, deep wrinkles, and yellowing teeth, and is wearing an off-white shirt that looks too small for her frame. She reeks of cigarette smoke and is talking so loudly that I'm positive I'm not the only one who can hear every word she says. Standing next to her is my waitress, her facial expression clearly telegraphing that she would rather be anywhere else than listening to this woman blather on. The older waitress obviously doesn't know how to read body language because she doesn't stop talking.

"Can you believe what happened on that ship? This is why I'll never go on a cruise! I can't imagine what those people went through. How in the world does something like that even happen in this day and age?" she asks.

You'd be hard-pressed to find anyone who hadn't heard the story. When the story first broke it didn't matter where you were —at the local coffee shop, grocery store checkout lines, and bank lines—you could overhear someone talking about it. Every magazine and newspaper cover featured a photo of the ship, accompanied by bold, eye-catching headlines. People were obsessed, but who could blame them? Considering the photo that was shared everywhere, it's only natural that everyone would be talking about it. Hell, even I was blown away by the image.

It was like witnessing the aftermath of a bad car accident on the highway. You want to see what happened, who was hurt, and what the damage was to the car. Some part of you wants to see the blood and broken bones, even if you'll never admit it to anyone. Out loud, you might say things like "I hope everyone is okay" or "I wonder what happened," even though, deep down, you secretly want to see the worst possible thing. Naturally, people are curious about strange things, and your inquisitive nature pushes you to learn more. Just like the car accident that gave you an adrenaline rush or a train wreck making local headlines, the "Parrish Cruise Lines Disaster" had a similar effect, and the entire world stopped to watch.

How could you not be obsessed? There were 2,100 passengers on board, including 700 crew members and 1,400 of their families and friends. This ship was the first to be built for the start-up cruise line. It was to be the flagship of the many vessels planned for the new company. The trip served as a training exercise to ready both the crew and the ship for its maiden voyage.

The itinerary was four days long, with two days at sea and one stop in between. The main purpose of the voyage was to test all the departments, from the service elements, to show performance and the dining experience.

The story started when an unknown blogger posted a single crime scene photo online. No one knows how they were able to get the photo; some said the police leaked it, but nothing has been proven. The photo was of the atrium. In the picture, you could see the once beautiful wrap-around glass staircase that reached across three floors, with most of the steps shattered into hundreds of jagged pieces and covered in a thick, almost inky-black substance that had dried and cracked in the same way a lake would during a drought.

The wooden railing that once followed along the stairwell had been snapped into pieces, leaving a large gap between the floors. Halfway down, you could see the remaining portion of the broken railing sticking up from the atrium floor like a spike. At the base of the stairs rested a decayed corpse, the jagged wood railing pierced directly through its torso. On the right side of the atrium were two large elevators encased in tarnished gold. Their observation windows were shattered, and the cars were demolished beyond repair, crumpled like crushed cans, as if they'd been hurtled down to the bottom floor.

Above the atrium, a corpse was hanging by a rope from an enormous, lopsided crystal chandelier. Around the neck of the corpse was a white lifebuoy with the word "TRAITOR" written on it in what looked like dried blood.

A pile of bones sat on the floor atop a white sheet, directly under the hanging corpse. The bones appeared disturbingly clean, completely stripped of all muscles and flesh. The way they were left looked like they were discarded after... something happened. Something unspeakable. Something violent and desperate.

Most of the entrances leading to the atrium had been completely blocked off with furniture. The only way to enter was to jump from the balcony into the shattered glass stairwell or through a single entrance on the 4th floor.

Everything you've read in the media was based off that photo. No one truly knows what happened.

No one knows the utter descent into darkness and depravity that happened at sea for nine nightmarish months.

Well, that's not entirely true. One person knows everything. Every sordid, fucked-up detail. How do I know that?

Because I was there.

PART 1

JUNE 2012

1

THEO

As we board the ship, I immediately turn to see my wife Julie's reaction as she crosses the gangway.

"Aren't you amazed?" I ask with a broad smile. She hesitates before responding, and I observe her carefully as she takes in the opulent interior. Currently, we're the only ones in the atrium, and I can't help but notice how stunning she looks in the gentle glow of the chandelier's light. Her eyes widen, and a smile as broad as mine spreads across her face.

"It's incredible," she says softly. "I'm so proud of you. It feels surreal to be here after all the years of listening to your dad talk about building this ship."

I didn't expect her to bring up my father. I feel a tinge of sadness at the thought, but I push through and keep smiling. "You're right, babe. It has been a long time in the making. I feel like I've worked my entire life for this moment. I wish my dad were here to see it with us." My smile must not be all that convincing because Julie quickly grabs me by the hand and changes the subject.

"Hey, how about you show me around before the passengers start to get on."

Quickly, my thoughts turn positive. "Right this way, my lady," I say, placing my hand on the small of her back to nudge her forward.

As the owner and first officer, there are perks like getting to board before anyone else. I know I'm not supposed to let my wife or anyone else on board before everyone else, but I have some pull, and security turned a blind eye. Before we arrived, I mentioned to my brother, Tyler, that he could join Julie and me for an early tour of the ship if he wanted to. However, when it was time for the tour, as expected, Tyler never showed up, so we decided to start without him.

This is the first time Julie has seen the ship since all the furniture and interior touches were finished. Six months ago, Julie and I had the chance to walk through the ship while it was being built, but at that time, only the structure was complete, and none of the interior decorations or furniture had been installed.

"I could never have imagined the finished ship would look so amazing," says Julie as she slowly takes in our surroundings as we walk. I agree. Having been on the ship multiple times, learning every room, practicing all the emergency drills, and becoming familiar with all the equipment needed to navigate, I never once stopped to admire the interior. Feeling every ounce of pride and joy, I take a moment and look around at all the beauty the ship has to offer. I know that in just a few moments, the peace and quiet will be over, and this ship will be bustling with excited guests.

As soon as you step on board, you're greeted by a breathtaking three-story atrium. The polished pearl-white floors feature various sea creature designs, creating the illusion of ocean life swimming at your feet. At the center of the room, a stunning glass staircase winds its way up three floors, with frosted glass steps that cast a soft white glow. The mahogany wood railing is stained a rich reddish-brown and carved to resemble intertwined ropes. To the right of the atrium, you'll find two elevators with glass walls and gold-accented doors. A chandelier with 500 teardrop-shaped crystals arranged in a spiral hangs from the ceiling, emitting a gradient of light blue to sea green from top to bottom, mirroring the colors of the ocean. On the left side of the atrium, the guest services desk is just as luxurious, featuring quartz countertops lit from below with sea creature designs etched along the edges.

After standing in the center of the room for a few moments, I feel Julie tug on my arm.

"Let's go," she says, excitedly. "I'm dying to see the rest. I'm so excited to check out the pool deck and do some sunbathing." Together, we rush to the elevators, smiles plastered on our faces like high schoolers on their first date.

Entering the elevator, I press the gold button labeled "12." It slowly ascends. When the doors open, we step out into a warm breeze that ruffles our hair and soothes our skin, struck by the stunning view. The 12th deck is quiet, with only a few crew members preparing for arriving passengers. As we stand there, enjoying the breeze, I take Julie's hand and gently tuck her strawberry-blonde hair behind her ear. I lean in for a deep kiss, but she pulls back briefly before smiling.

"Save that for later, mister. I want to see where you'll be working."

"Okay, okay, it's this way."

I take Julie by the hand, and we walk up to the 13th floor and the bridge. Inside, several seats and screens line the entire length of the room. I make my way over to where I'll sit when working, grab the chair, and swivel it around. "Come on, take a seat."

Julie sits down and turns to face the screen. I quickly pull out my phone and snap a photo of the moment. "Well, what do you think?" I ask, smiling. She smiles back and lets out a laugh.

"What are you laughing at?" I ask, confused.

"This isn't what I was picturing at all. Where's the big wooden steering wheel like on TV?"

"This is a cruise ship, honey, not a pirate ship and it's a lot different than the others I used to work on." Laughing, I point to the joystick in front of her and do my best to explain. "This is how we steer the ship. There's a small steering wheel we use for tight maneuvering or docking, but it's not like in the movies."

She nods, and I can tell right away she's not really getting how a joystick can steer a large ship. After years of being married, I know she really isn't interested in the logistics, so I move on and decide to show her some more restricted areas, hoping to impress her. We walk back out into the sun and feel the warmth on our skin. Before we continue, I grab Julie by the waist and pull her close.

"Thank you for being here with me. Of all the people in the world, I am so happy to share this incredible moment with you, my beautiful wife." She gently cups my face and pulls me in closer, locking us in a slow, passionate kiss. My heart begins to pound, and I feel a twitch between my legs as she runs her palm along the crotch of my white slacks. Instantly, I feel myself growing at her touch. I look down and realize that my bulge is extremely obvious. "Hey, don't get me too excited; I can't walk around with this."

She giggles, then leans closer to my ear and whispers, "Then show me our room," giving me a quick squeeze, and I almost lose it right there on the 13th deck. Excited and eager, I look around and check my watch.

"Okay, I wanted to wait until later to show you the room, but now works great too," I say with a wink. I'd specifically selected the room we would stay in—the presidential suite.

Another perk, of course. Grabbing Julie's hand, we walk aft of the ship where, behind the elevators, a small hallway is lined with the ship's most luxurious rooms. At the end of the hallway, a single door faces us.

"Close your eyes," I whisper softly, standing behind her and covering her eyes. "And no peeking, okay?"

"I promise I won't peek. But don't make me run into a wall. You know I'm clumsy," she responds.

"Babe, I've got you. I'm always here to protect you."

Julie stops in her tracks just as I reach for the door.

"What's wrong?" I ask. She looks back at me, her eyes filling with tears as she gazes into mine. She studies my features, and I can't help but study her right back—her smooth skin and bright ocean-blue eyes looking back into mine. She runs her fingers through my thick, jet-black hair. I can tell she is thinking about something, but I've got no idea what. Then a tear rolls down her face.

"It's nothing, it's just..."

I pull her in closer. "I mean it. I love you, and I'll always protect you," I say, wiping the tear from her cheek and pulling her in. Things have been rocky between us but I want her to know how serious I am. Between my father's illness and subsequent death, and getting ready for this maiden voyage, I've barely been around and the cracks in our marriage are starting to show. I'm hopeful this trip is exactly what we need to heal us and get us back to where we used to be. When our lips meet, sparks fly, reminding me of our wedding night so many years ago. I reach behind her and open the door.

The smell of roses surrounds us both. She turns around to see the suite floor lined with red rose petals leading to a room in the back. As she walks toward the bed, I notice her gasp slightly at the sight. The white marble flooring is so polished that you can practically see your reflection. To the left, there's a white grand piano with its massive lid open, revealing the interior of the delicate instrument. To the right, a full bar with shelves lined with every type of liquor imaginable. She continues to walk, following the rose petals to the master bedroom. Her eyes go wide when she sees the king-sized bed completely covered with more red rose petals. Behind the bed is a long teal curtain framing a massive floor-to-ceiling window. I grab the remote from the bed and push the button to open the blinds, and the curtains begin to move, giving us both a view of the open water.

From the look on her face, I can tell she is mesmerized by everything, exactly what I wanted. Since the ship is still docked, Julie grabs the remote from my hand and pushes the button to close the curtains. I watch as she glances at the nightstand, where a large crystal vase filled with two dozen red roses sits alongside a platter of white and dark chocolate-dipped strawberries. Then she turns her attention back to me, eyes locked on mine. There is a hunger in her eyes that makes my body nearly explode. She walks toward me, turns around, and lifts her strawberry-blonde hair so I can unfasten the clasp of her dress. With a quick motion, I unclip it and slowly slide the dress off her shoulders, revealing her perfectly pale skin.

As I run my fingers down her arm, I feel her body shiver. She slowly turns to face me. I stare at her for a moment, taking in her beauty, then gently lead her backward to the bed, our lips locked as she unbuttons my shirt. I struggle to pull off my belt; I swear I can't get it off fast enough. After a moment, we are both naked, and I lay her back onto the bed.

Silky rose petals press into her skin, and she smiles at me, her lips swollen and red from my nibbling. I climb on top of her and caress her breast, kissing her neck. Her body shivers at my touch, letting me know she is enjoying this as much as I am. With every touch, I feel her quiver and moan. Intertwined, we make love like it's the first time. When we finish, I stare into her beautiful blue eyes and gently slide her hair behind her ear. "I love you so much."

"I love you too," she says, kissing me once more, then sitting up and grinning at me. She heads naked into the bathroom, leaving me on the bed. I let out a sigh and slump back onto the bed. God, I love that woman. *May this trip be exactly what we need,* I wish as I stare after her.

After we've both showered and changed, me into my captain's gear and her into her hot-as-fuck bikini, we head to the pool deck. Once there, we sit down at the bar. Since I'm working and can't have alcohol, I order some water. Julie, on the other hand, is ready to party right away.

"Can I get a frozen margarita without salt, please?" she asks the bartender, who's setting up the bar for the guests arriving soon. I lean in and give her a kiss.

"I'm going to track down Tyler and try to catch up with you all before dinner." She rolls her eyes and waves me off as she waits for her margarita.

I know my brother has some issues, but I'm not sure why Julie makes such a big deal out of it. Sure, Tyler drinks a bit too much, and I stay out late when he's around, but that's nothing new. He's always been like that, and I doubt he'll change now. I try to ignore his problems and not let them get to me. I can't force Tyler to find a job or stop relying on the inheritance from our dad. I know the money will eventually run out, but Tyler is stubborn and refuses to listen to anyone else. Even if Julie isn't fond of him, he's still my brother, and she is my wife, so they'll need to find a way to make it work.

Tyler is a laid-back guy who's always the life of the party—which is fitting, since he's always out and about. Tyler and I might share the same birthday and even physical features, but despite being identical twins, we have completely opposite personalities. Maybe that's why I enjoy having him around so much. I glance down at my phone, checking to see if Tyler has replied to any of the dozen texts I've sent since last night. Nothing. "What the fuck, Ty?" I sigh and shake my head. Whatever is going on with him, I don't have time for right now. I head toward the control room. I have a ton of shit to do, and I'll have to deal with Tyler later.

2

TYLER

"Fuck!" I shout, stumbling out of bed and crashing to the floor, hitting my knee in the process.

Assuming the hotel clock is right, I have only a few hours until the ship departs. Did I sleep through my alarm? Hell, did I even set one? My alcohol-soaked memories of last night are hazy at best. I throw on my clothes, brush my teeth, and quickly check my phone. Ignoring the several missed calls and dozens of texts from my brother, I open the ride-share app and order a ride, hoping traffic isn't too bad. Last night must have been wild because I can't remember what happened or who the brunette who's still half-asleep on the mattress is. My head is throbbing from the hangover, so I don't try to recall her name.

"Hey," I say to the woman. "Get up, you need to leave. I'm late." I pull back the curtains, letting in the bright sunlight that worsens my headache. I squint and face her again. Noticing she isn't moving. "Really, get up—it's time to go."

She rolls over and groans in bed, then responds hoarsely, rubbing her head. "Chill out, dude, I'm getting up. And can you please close the blinds? My head is pounding."

"Never mind, stay in bed. But check-out is in an hour. I need you to leave before then," I tell her, annoyed.

"Okay," she says, flopping back onto the pillow and covering her face with the duvet.

I slam the door behind me and rush down to the lobby, nearly tripping over my untied shoes. My phone vibrates in my pocket, signaling the driver's arrival. After confirming the license plate, I hop inside and ask the driver to go as fast as possible. Within minutes, my phone buzzes again—another message from my brother, Theo.

Theo: Dude, where the hell are you?

Theo: Why are you ignoring me?

Theo: Hello?

Theo: You better not mess this up or else! This is a big deal for me. Get your ass over here now because I'll fucking leave you behind!

Theo: Seriously, where are you?

Theo: Why is your location off on your phone? I see you're getting these messages. You better not be ditching me.

Theo: I ALSO SEE THAT YOU'VE READ MY MESSAGES!

Me: Calm down, bro!

Me: I'm on my way. I overslept. I'll be there, relax.

I ignore the rest of Theo's nonstop messages. My head is still pounding. Reaching into my bag, I feel around until my hand lands on a pill bottle. Opening it, I take out some ibuprofen, swallow them dry, and pray that they kick in soon. Glancing at my watch, I ask the driver to wake me when we arrive. Fifty minutes later, the driver is hollering at me from the front seat that we've arrived. Quickly, I gather all my bags and get out of the car. Even though my dad built the ship and my brother now owns it, I still have to go through security just like everyone else.

At last, I'm crossing the gangway and boarding the ship, albeit a bit late. The ship is buzzing with activity, loud and packed with guests. The ibuprofen is finally starting to take effect, so the noise doesn't bother me as much as it would have earlier.

I finally take a second to look around, and holy shit, it's amazing. I knew the ship was nice, thanks to Dad's constant chatter about it and his tales of how much he'd invested in the cruise line over the years. But I was unprepared for how impressive or how large it is. I scan the atrium, trying to spot my brother or Julie, my sister-in-law, but all I see is a crowd of unfamiliar faces chatting away.

My phone vibrates against my leg, and I pull it out to see Theo's name flashing on the brightly lit screen.

Theo: Look up.

Me: Where?

Theo: Your right side.

Theo: Your other right side, dumbass.

When I look up, I spot my twin brother in his all-white uniform, mingling with guests and shaking hands. I walk up the glass stairwell, taking in the luxurious surroundings. My dad spared no expense making this ship opulent. I also notice how much it has changed since my last visit — back then, it was merely steel frames and scaffolding. Dad had insisted his sons see it, claiming it would change the world.

"Tyler, over here," shouts Theo, waving a hand to get my attention.

"Damn, bro, I look great all dressed up," I respond, grinning.

"We clean up pretty well, don't we? Rough night?" says Theo as the group of guests he was chatting with says goodbye and walks away.

"Why do you ask that?"

"Because you missed the early tour and ignored all my texts."

"Oh, yeah, I had a good time and forgot to set my alarm," I say, nonchalantly, as if it's a regular thing, which it's not.

"Well, I hope you're ready to keep the good time going," says Theo, grinning from ear to ear. "Everything's included, and you're in one of the nicest rooms on the ship, aside from mine, of course."

"Hell yeah, bro, I'm ready," I say, looking around. "Where's the bar? I need a drink ASAP!"

"Well, Julie is at the bar on the pool deck if you want to go say hi and get yourself something. Have one for me while you're at it, since I'm, you know, working. Something you should try sometime," says Theo, laughing. I don't see the humor.

"Oh, fuck off. I don't need a job right now. Let me live my life, bro. I'm good on the meetup with Julie. You know that woman can't stand being around me."

"I don't have time for you two going after each other," Theo says, almost under his breath. He shakes his head. "Look, each member was allowed to choose two guests for this test voyage. I chose you and Julie. Can you at least try and get along, and not embarrass me, please?"

Instantly, I get aggravated by that, like he is blaming everything on me, as he always does. "Well, I hope you told her the same thing. I'm not the one who usually makes the rude comments."

"Please, Tyler, can you just give it a try for me?"

"Fine, I'll go find her," I say, knowing he genuinely is asking me this time. *Here's hoping I don't regret my decision.*

3

JULIE

"What do you want, Tyler?" I ask, clearly annoyed, as Tyler pulls out the barstool next to me.

"This was the only seat available," he shoots back, his tone as sarcastic as mine. I glance around the pool deck, which has bars on both sides, and notice plenty of empty seats at each.

Instantly, I turn back to Tyler and point to the empty chairs across the deck. "What's wrong with all those other seats?"

"I didn't see them, I guess. Oh well, I'm here now," says Tyler with a wide grin. I give him a death glare, imagining how much I'd love to slap that smirk off his annoying pretty face. One that looks exactly like my husband's, but at the same time not, due to his irritating personality. Tyler watches me, and I'm pretty sure he knows exactly what I'm thinking because he gives me a half smile. He slides his chair closer to me, knowing it'll only annoy me more. "Your generous husband said we should try to get along, so I came up here for a drink to make it happen."

"Since when do you listen to anyone, let alone your own brother?" I snap.

"There's a first time for everything. You, of all people, know how much this means to him, so let's set our differences aside. You can hate me again when we're back on land. Sound good?" he deadpans, then waves down the bartender.

I take a deep breath before responding, "Okay, deal. For Theo's sake, I can make it work. Just don't annoy the hell out of me, and don't get completely wasted and make a fool of yourself like you usually do." I smile and take a sip of my ice-cold drink.

"No promises, but I'll try." He responds with a quick wink and a smile. *Damn, he gets on my goddamned nerves.*

We both sit at the bar, silent for a moment, sipping our drinks and actively avoiding each other, until my husband walks over, surprising us both. This drink is so strong, I already have a buzz.

"You look sexy as hell in that uniform," I say, licking my lips and imagining what he did to me earlier.

"Now, look who's the one making a fool of themselves," Tyler says, with a snort.

I can't help but roll my eyes every time he speaks. "Shut up, you asshole. He does look good."

"Yeah, shut up, dude. I love it when she tells me how sexy I am," Theo says, winking at me. Then he leans down and gives me a kiss. I feel heat inside my body again at his touch.

Tyler smirks, and before he opens his mouth to speak, I'm already preparing for something dumb to come out. "So you think I'm sexy, too, right? Since we look exactly alike?"

And there it is.

"You really know how to kill a buzz, Tyler," I say, rolling my eyes again.

He lifts his eyebrows. "But you didn't deny it."

"No, maybe you would be hot if you didn't have the personality of a high school jock." I scoff and turn my attention back to my drink, so I don't punch him in the teeth.

"Just keep in mind that if Theo ever leaves you, I'll still be here for you, and you won't even need to change your family photos. Just tell everyone it's me in those pictures." I look at him again, wishing I hadn't, because he has the same smirk as before. The one I can't stand.

"Luckily, I'm not leaving, and she isn't interested in you. I came to remind you both that the sail-away party is about to begin, and I need to head to the bridge to work. I'll see you both at dinner. Love you, have fun!" Theo says before kissing me on the forehead and heading toward the bridge.

Tyler sits there bugging me with his existence for another twenty minutes before he heads to his room, leaving me alone at the bar—thankfully. I look around and realize people are starting to arrive and take up all the lounge chairs. Holding my third margarita, I get up and walk around in search of a lounge chair to relax in. Enjoying the hot sun and warm breeze on my skin, I close my eyes for a brief moment, allowing the sun's warmth and the sounds of the deck area to buzz around me. It's been a long time since I've had a few moments to myself, and I want to enjoy it.

Soon, the pool deck is packed with excited passengers. The music blares, and people dance and celebrate. Amid all the noise, I choose to watch instead of joining in. Lost in my own thoughts, I'm suddenly jolted back to reality when a group of people jump into the pool, splashing water onto my legs. It's refreshing, and I think about jumping in myself, then I think better of it. Meanwhile, the DJ spins old-school dance tunes that energize everyone. The beat is contagious, and I find myself tapping my foot and nodding along.

The pool is now overly crowded, and a long line has formed for the waterslide. For a moment, I think about getting on the slide before dinner. Then, I immediately dismiss it. I have no desire to ride that plastic deathtrap. It's not a typical water slide where you sit in a raft and glide into the pool. This one is a bright blue and green drop slide that starts inside a dark tunnel. After stepping in, you cross your arms and legs while waiting for the floor to suddenly drop from beneath your feet. You plunge down into a steep spiral before the slide turns transparent and extends about 15 feet over the ship's side, revealing the ocean far below. The entire design is meant to create a thrilling rush of adrenaline, and I remember Theo's dad talking about how excited he was when he chose that design.

When the ship's horn sounds, the crowd claps, dances, and cheers, letting me know that the ship has departed from the port and is now underway. Trying to escape the noise and chaos of people dancing and splashing in the pool, I head toward the elevator. A blue sign beside the elevator doors shows an outline of the ship with a red dot on the 12th floor, along with the words '*You are here.*' Below the map is a list of all the areas, each color-coded and shaded according to the outline. Standing there, I'm unsure where I want to go. Checking my watch, I note that I have a few hours until dinner. Deciding to find a quieter spot to unwind, I study the ship's layout and spot the library on the 6th floor. I pull out my phone, thankful I still have service, and text Theo.

Me: Hey honey, I'll be in the library for a while, dying to get into a book. I can't believe that in all the packing I forgot to pack mine. I'll see you at dinner. Good luck today. Xoxo I love you.

Since I don't get a reply, I assume he's busy. Still holding my phone, I find Tyler's contact and message him, telling him where I'll be in case he needs me. Although, honestly, it's none of his business, but I did agree to cooperate with Theo's request to get along.

Me: Heading to the library to enjoy the quiet. If you need anything, *don't* need anything.

Tyler:

I'm surprised that Tyler responded. Usually, he ignores

everything I say. Maybe he's trying, like Theo asked. If Tyler can be civil, I suppose it's the least I can do, too. Although if Tyler's behavior is anything like it's been in the past, God only knows what he's up to or with whom. Hell, he's probably at the bar, looking for some hapless woman to seduce for the next few days on the ship. I'm relieved to know Theo is working the evening shift and not out partying with his brother. Since it's only a training cruise, Theo gets to switch out with other crew members. Luckily, that means he will be off for the remaining three days after today. With nothing to stress about right now, I'm excited about having some me-time.

As the elevator comes to a stop on the 6th floor, I can't help but gaze through the glass walls, taking in the bird's-eye view of the atrium. From inside the elevator, I can see that the party on the pool deck isn't the only place with excitement. The atrium is just as lively, with its own unique energy. The elevator's recorded voice softly announces the 6th floor, and the gold doors slide open. Stepping off, I walk over to the railing and look down at the guests. The crowd in the atrium is a world apart from the one on the pool deck. Upstairs, the crowd is younger and rowdier, with lots of drinking, yelling, and jumping into the pool. The pool deck's DJs are playing music that makes you want to get up and dance.

In the atrium, the atmosphere is much more subdued. The crowd is older, with the women dressed in upscale, semi-formal attire, and the men in business casual. A man in a sleek black-and-white tuxedo is playing the piano, accompanied by a woman in a stunning full-length gown with emerald-green sequins and deep red lipstick. It's hard not to marvel at her beauty. Her brunette hair is pulled back tightly with a gorgeous diamond-covered barrette. She stands onstage, chin-level with a 1920s-style microphone, singing slow, soulful versions of popular songs. Right now, she is singing "Someone Like You" by Adele.

Before I forget why I came here in the first place, I turn my attention towards the library entrance and make my way inside, where it's so quiet that I catch myself still humming to the song. Looking around, I realize I'm alone, grateful to have the place to myself. After walking around the different sections, I stop in front of the *Romance* section. Without hesitation, I reach for *The Last Song* by Nicholas Sparks. It's a comfort read for me and ranks among my top five books by him. I take the book to the large, round porthole window, which has a big bench with a soft and comfy cushion on it. Sitting down, I take in the stunning view—the cruise port is now in the distance, gradually fading away with each passing second. The white caps of small waves break across the rough water as far as I can see. Taking a deep breath, I open the book and escape to my happy place.

4

THEO

Standing on the bridge, I gaze out through the large glass windows.

"Steady, keep it steady. You've got this. You can all do this in your sleep by now," shouts Captain Mark Whitaker.

I hope Whitaker can't see how nervous I am right now. I know Whitaker is right; I *can* do this in my sleep. But this time feels different. This time, I'm in control of the ship my dad spent years building. Since this is my first time navigating this ship, I'm extra cautious; I don't want to mess anything up. Whitaker walks over to me and places his hands on my shoulders.

"The day has finally arrived. Be proud of yourself. You've worked so hard for this, and I hope to see you as a captain someday. You have real potential, my boy," he tells me. I take a deep breath and smile. Just like that, all my nerves fade away.

"Do you really mean that?" I ask, uncertain.

"I wouldn't say it if I didn't mean it, son," Whitaker says, giving my shoulder a squeeze. "Your father and I were friends well before you and your brother were born, and I've had the pleasure of watching you both grow into the men you've become today." He lets out a deep breath, then puts his hands into his pockets. "It's a real shame he's not here with us today. He told everyone he knew how proud he was of you. Often."

I clear my throat, keeping my eyes on the open water ahead of me. "Thank you," I say quietly. I blink a few times as I quickly wipe away a stray tear. "We all thought we had more time." Because he was so ill, Whitaker and I moved the date of our sailing forward several months and even arranged for a team of doctors and nurses to be on board to care for my dad so he could join us. But it was still too late.

Whitaker doesn't respond; he just stands there, looking out at the water, remembering his old friend, as I speak up. "You know, after he died...I almost quit."

After a moment, Whitaker replies, "I didn't know that."

I look over at him, the knot in my chest squeezing a bit. "This was supposed to be *his* maiden voyage, not mine. Julie convinced me to take it."

"Smart girl, that Julie. We all know this ship meant everything to him. He poured his heart and soul into building this company. Besides you and Tyler, this ship meant everything to him. So yes, Theo, you are meant to be here today, living out the dream for your father. And I want you to know I'm proud of you, too. I'm glad you're here." He grins at me, giving me a firm pat on the shoulder. "But, for now, I need you to focus and pay attention. You have a ship to command. Crashing it on the first day out would be terrible for our Yelp reviews," he says with a laugh as he heads aft of the bridge.

I chuckle as I focus my eyes back on the water. I'm glad he told me that. Whitaker is more like family than my boss. Outside of work, Tyler and I call him Uncle Mark. When we were ten, a drunk driver hit our mom, and she didn't survive.

Mark was our dad's best friend and was there during our dad's lowest moments. That's when his idea of starting a cruise line was born. Growing up, we always went on cruises, which the whole family loved. Mark worked for a cruise line then as well, and often invited our family to join him whenever we could. Our dad loved being on the water and cruising so much that he told Mark he'd someday build a cruise business from scratch and wanted Mark to be the captain. Coming from a wealthy family, our dad had the means to make it happen, and he meant every word he said. I loved the water as much as my dad did. Given that he'd been my only living parent, everything he said meant the world to me, and I wanted to make him proud.

With my father being a widower, Tyler and I inherited the family company, Parrish Cruises. Tyler made it clear right away that he wanted nothing to do with owning the company. He and our dad didn't have a good relationship, so Tyler signed everything over to me as soon as he could, giving me full ownership. I didn't know much about running a company, so I hired people who did. Parrish Cruises has an entire finance department, giving me the freedom to work on the ship as I always wanted, without the burden of everyday business responsibilities.

"Theo! Get your head out of the clouds," shouts Logan, standing right next to me.

I startle, giving him a death stare for screaming in my ear, and he laughs. I fight the urge to punch him in the throat. Logan's been giving me a hard time since the day we met. His dad was one of my dad's childhood friends, so that's how he got the job. Logan is older with a military background, and has a higher rank than I do on the ship—something he never lets me forget. Thankfully he's good at his job, which is his only true redeeming feature in my mind. He treats me like a child, especially when the captain steps away, leaving Logan in charge.

When that happens, Logan does everything he can to make sure that I know he is the one calling the shots.

"Did you hear me?" shouts Logan again, flopping down into the captain's chair.

"No, you didn't say it loud enough. What did you say?" I ask.

"It's almost time for dinner, and since you're taking the first break, I need you to make sure everything is in order before you leave. Go over the checklist," he demands, his tone sharp.

I ignore the attitude, refusing to let him ruin my evening. "Yeah, I got it covered. I'll do it right now. Sir." I fight back the urge to tell him he's only in command because I allow it. But I bite my tongue and force a fake grin.

"Thank you. Have a good dinner. Say hello to Julie and Tyler for me. And don't be gone forever, I'm starving," he says with a grin, sitting back in the chair, propping his feet on the counter. Anyone else would think he's joking, but I know he's never missed a free meal in his life. If he isn't eating, he's working out.

"Where's the checklist?" I ask teasingly, knowing the question will irritate him more.

"In the same spot as always. Damn, Theo, do I have to do everything for you?"

I laugh at his predictable irritation. "Relax, it was only a joke. I know where it is." Logan stays silent, but the annoyed look on his face tells me I've pushed his buttons, which is precisely what I wanted. I walk toward the cabinet with the emergency gear and find the binder labeled 'PROTOCOLS'. I flip to the page with a checklist of all the emergency equipment and walk around the bridge, checking each item to ensure nothing is missing. Hm, that's weird. "Logan, do you know if the two survival rafts were moved? They're not here."

"Dude, that's not my job to know. Figure it out." I clench my fist, ignoring his smart ass remarks and walk away to finish the checklist. I swear they were ordered with all the other emergency equipment, but I have no idea where they could be. I try not to worry—worse comes to worst, I'll order more after this trip. There are plenty of lifeboats on the ship if needed.

I continue down the list. There are only a few more items to check off, then I can get some food. The process takes me about ten minutes. I stop at the last item on the list, which is to verify that the controls for the doors and locks are functioning properly.

"Have you seen this?" I ask Logan while staring at the control panel's screen, which flashes a large 'ERROR' message in bright red.

"Have I seen what?" he asks. I don't even know why I ask him. He is no help, just an asshole.

"The error message on the control panel for floors 2, 3, 7, and 8."

"Oh yeah, I saw that. Whitaker said it's okay. Something to do with the override used to block off the unfinished floors."

"Okay, well, I'm going to check it out and make sure everything is still locked. I still have some time before dinner, and I haven't seen what those floors look like. I'm curious," I say, staring at him.

"Go ahead, superstar. You're wasting your time. I told you, Whitaker said it's all good, so it's good." His sarcastic tone sets my nerves on fire.

As I head toward the door, I knock Logan's feet off the counter, causing them to slam to the floor. I look Logan in the eye. "You might be second in command, but don't forget who signs your paycheck."

I can't help but grin as the last thing I hear before the door shuts is him grumbling, "Asshole."

As I walk to the dining room, I think back to a decision the board of directors and I made a few months ago, unanimously deciding to move the scheduled training date up three months to accommodate Dad. At that time, his health was rapidly declining due to cancer, and we all wanted him to sail on his ship before he passed.

Unfortunately, he didn't make it, but after his passing, the plans to sail early were already in place. I also found out my dad dished out a shit ton of money to falsify the paperwork to let the ship sail with unfinished floors and keep it under the radar. I was told multiple times that moving the date back to the original date would only cause chaos and bring attention to the fact that we were never approved to actually be sailing— officially... no one, including the Coast Guard, knows we are sailing. The Port Authority was paid to look the other way, as were the inspectors.

The design company building the ship assured all of us that the unfinished floors wouldn't cause any issues during the cruise, as long as they were blocked off for the entire sailing. The elevators would be programmed to bypass those floors, and only the highest-ranking employees could access them. The bright red error message on the control panel makes me wonder whether we made the right decision. The last thing I need right now is someone gaining access to those floors and getting hurt. I can already imagine the wave of lawsuits.

Stepping into the elevator, I press the gold button engraved with the number 8. The light around the button blinks red, indicating I can't access that floor, which is a good sign. This means the error message isn't related to the elevator's access to the blocked floors. I pull out my key card –which also serves as my company ID– and place it on the black box above the elevator's keypad, granting me access to all floors. I press the 8th floor button again. This time, the light flashes green, and the elevator begins moving. My keycard allows priority access to the elevators and bypasses all the other floors, preventing anyone else from entering. After a few moments, the elevator stops, and the door opens on the 8th floor. I blink for a second, in complete shock at what is in front of me.

"Oh— my— god."

5

TYLER

Sipping my glass of champagne, I scan the room for an empty seat. With some downtime, I'm looking to unwind a bit. I spot a group of women standing and talking among themselves, giving me the perfect opportunity. Making my way toward them, I catch the eye of a younger brunette. I smile. *It's time to turn on the charm.*

"Would you mind if I sat here?" I ask, motioning toward the seats they're blocking.

The woman smiles back at me. "Go ahead. Sorry if we were in your way."

"I wanted this spot for the view," I say, staring directly at the brunette.

She gazes at the stage, taking in the scene. "I'm obsessed with that gorgeous green dress she's wearing, and her voice is amazing."

"That's a nice view, too, but I wasn't talking about her. I was talking about you."

"Oh." Her cheeks flush pink, and she giggles.

"My name's Tyler. What's yours?" I lift my glass to my lips and take another sip.

"I'm Kelly, nice to meet you," she says, reaching out to shake my hand.

I set the glass down on the nearby marble table and take her hand, gently bringing it to my lips and kissing the back of it. "The pleasure's all mine."

Kelly wears a sleek black dress that hugs her curves and highlights her toned legs. She's in great shape. Her brunette hair is swept to one side, cascading over her shoulder. As she talks to me, I can't help but notice her licking her lips, which are a glossy shade of red. Then she runs her perfectly manicured fingers through her hair. I feel myself harden at the sight. I can't help but imagine my hands entwined in her long hair, pulling it as she screams out my name. Lost in a daze, I don't hear her ask me a question. But I notice she is no longer talking and is staring at me, waiting.

"I'm sorry, what did you say?" I ask. *Shit. I really hope I didn't just fuck this up.*

"I asked if you were here with anyone?"

"Yeah, I'm here with my brother and his wife. What about you?" I ask, doing my best to subtly adjust my growing bulge without her noticing.

Kelly suddenly rises on her tiptoes to peer around me. "I'm with my best friend, Sara, who's dating one of the cooks. I think his name is Chris. They invited me to join them. She was here a few minutes ago."

"That's nice. Do you have any plans for tonight?" I ask, not giving a damn about her friends.

Kelly, still scanning the room, responds, "Not really. Sara and I were supposed to hang out and grab drinks, but it looks like she's stood me up."

"What about those ladies?" I point to the group of women she was standing with. Not that I care, but whatever.

Kelly waves her hand to dismiss them. "Oh, they all know

each other, and one of them knows Sara. I was only standing with them, so I didn't look ridiculous in a corner all alone."

"I'm free tonight if you'd like to grab a drink," I say with a sly smile.

"I already have a drink." She smiles, holding up her glass.

"I meant another one after this. We can find a less crowded place and get to know each other better."

I sense she knows exactly what I mean because she responds quickly, licking her beautiful red lips. "Sounds good to me. I have no idea what those women are talking about anyway, and you seem way more interesting."

I motion toward the elevator. "Let's head to one of the lounges with a bar. I need something stronger than champagne. Maybe at the Stardust Lounge. I overheard someone say it's really nice." As we wait for the elevator on the atrium floor, I look up and spot Julie leaning over, watching everyone. A moment later, she's gone. Hoping Kelly doesn't mind, I tell her, "I'll meet you in the Stardust Lounge in a few. I just saw someone I need to talk to. I'll only be a minute."

"Okay, see you there," she says, walking away, and I can't take my eyes off her perfectly plump ass.

Instead of taking the elevator, I take the stairs up two floors. Julie's not standing at the rail anymore, but the library's glass doors are just off to the side. From what I can see, the library appears to be closed; there's no one inside browsing or sitting around. I open the door and look around just to be sure, but I don't see her. The library is surprisingly small, so there is no way I could have missed her. So where is she?

I navigate the bookshelves again, but I still don't find her. Just as I'm about to leave, I catch a glimpse of movement from the corner of my eye. I turn to look, and that's when I spot her lounging on a window seat in a large circular window. She doesn't see me, and I take advantage of that. Watching her carefully, I walk silently toward her, planning to startle her. But I stop short when I see her pull out folded pieces of paper and start writing on them. I can't quite make out what she's writing. Trying to get a better look, I walk slowly toward her doing my best to not make a sound. I notice the name "Theo" written at the top, but her hand blocks my view of anything else. As I get closer, I accidentally bump into the bookshelf and knock over a stack of books. "Fuck," I whisper. Julie quickly scrambles and shoves the papers away.

"What are you doing, Tyler?" she snaps.

"I can ask you the same question. What's that in your hand?"

"It's none of your business," she tells me sharply.

"Well, it didn't look like you were reading a book," I say with a smirk I know she hates.

"You know, if you weren't so irritating, people might actually like you. Your daddy's money won't last forever, and when it runs out, which it *will*, I don't want to see you come crying to your brother for help." She pauses for a moment, glaring at me. "He's worked his ass off to get where he is, and all you've done is party and drink your life away. You're not a kid anymore. Grow up, asshole." She rolls her eyes at me, shaking her head.

Stunned by her words, my grin fades. "Wow, that...okay. Ouch. That was harsh," I reply.

Julie takes a deep breath and some of the irritation in her eyes fades. "Okay, that was harsh, but I think you needed to hear it, Tyler. I'm writing a letter to Theo telling him how proud I am of him. If you weren't so damn self-centered, you'd see that he's been struggling. He needs you. You need to get your shit together; Lord knows he's always been there for you."

"I thought this was what he wanted: the perfect wife, the perfect job, the perfect life. How was I supposed to know he was having a hard time? He makes everything look so... easy." I run a hand through my hair. I had no idea shit had been so hard for him. "I'm sorry. I wasn't trying to be an asshole when I walked in here."

I can barely admit it to myself, but I'm jealous that he has it all together. I act the way I do because I feel lost. I sleep with a lot of girls and party all the time because it gets me attention; people act like they like me and want to be around me. My partying and flaunting my wealth hide the fact that I have no real talent, unlike Theo.

"All our dad cared about was this damn ship and Theo. He never had time for me or anything I liked. That's why, when he died, I signed over my share of the company to Theo. All this—the ship, the cruise line business—has always been their thing, not mine. I absolutely hate the ocean and boats. I figured if I just stayed out of Theo's way and let him do his thing, he'd be happy."

She closes her eyes briefly, then lets out a heavy sigh. "Tyler, you have no idea how much Theo cares about you. You mean so much to him. Why can't you see that? He doesn't want to drink and party every time he sees you. He does it because he thinks that's what *you* want. He just wants to hang out with you. He wants you to be happy. All you have to do is ask, and he'll gladly help. No one wants to see you fail. Use your money to do something good, or go back to school. When I first met you, you were trying to become a police officer. Whatever happened to that?"

"I didn't pass the initial training and was too embarrassed to tell anyone, and no one ever asked. All our dad ever cared about was the ship, which is why I hate it so much." I say now, my head down, feeling defeated.

"You should go back and try again. If it's something you love, do it for yourself, not for anyone else." Her tone is one I've never heard from her before.

"You're right." Everything she's said is valid, but it doesn't make it hurt less. I do need to get my shit together. "I... I'll work on it. I never meant to be such a shitty brother." I really didn't. Shit just... fell apart after Dad died, and I couldn't get back on track again.

"You're not a shitty brother," she says, settling back onto the window seat. "Besides, the first step in fixing your problem is admitting you have one. Go away now. I need to get back to writing." As she gets comfortable in the window seat, I notice a stack of books she almost kicks in the process.

"I didn't know you liked Nicholas Sparks," I say, staring at the books.

She glances at the books, then back at me. "Yeah, I'm a sucker for a sad romance."

"You can keep the romance books. That shit's boring. I'd rather make the romance real so a woman can think of me when she is reading about it." I wink at her.

"Get the fuck out of here, Tyler," she says, taking a flip-flop off and hurling it at me.

"I'll see you at dinner," I say, running out of the room before she can throw something else at me.

I make my way to the lounge to meet Kelly, at least I think that's what she said her name was. As I walk in, I spot her sitting at the bar right away. I walk over and pull out the chair next to her.

"Glad you didn't run away. Sorry about the detour."

"It's okay. I've got nothing better to do since my friend, Sara, ditched me."

I flag down the bartender and order a rum and Coke. I flash Kelly a charming smile. "Well, I have no plans other than drinking and getting to know you."

She licks her lips again and grins at me, then runs her hand up my thigh and whispers, "Well, I can think of a few things I want to do right now."

Feeling myself get hard for the second time in an hour, I'm glad to see that she has the same idea in mind.

"Oh, is that so? What do you have in mind, exactly?" I ask, licking my lips as I look back at her, staring into her golden-brown eyes.

She reaches down and runs her fingers over my growing bulge. "Let's go back to your room, and I can show you," she whispers, then plants a soft kiss on my lips.

I lean into the kiss, getting a taste of her cherry-flavored gloss, and I push the chair back and take her hand. "Follow me."

"What about your drink?" she asks as she follows me to the elevator.

"My brother owns the ship. I can get another one."

"Holy shit," Kelly says with a laugh. Together, we head upstairs to my room.

6

THEO

As soon as I step off the elevator, I immediately regret setting sail with the unfinished floors. When the board of directors approved sailing, we all assumed "unfinished" meant only missing furniture or minor details. This is far from what anyone imagined. I can't shake the thought that if a passenger accidentally accesses these floors, it could be deadly.

The 8th floor lacks nearly all the walls, most of the flooring is missing, and loose electrical wires hang everywhere. I make a mental note to verify that the power on the rest of these floors is off when I return to the bridge. The hallway appears to be only steel framework, with a long strand of lights along the ceiling, all visibly powered and brightly lit. Scaffolding is laid out for the construction crew, with empty cans and bottles left strewn on a few of the beams. Looking down, I spot the 7th floor, and the distance seems to be at least 15 feet. I can also see that it's partially completed.

As I look around, I feel the ship start to rock from the choppy water. The weather must be getting bad—I had spotted dark clouds on the horizon while I was on the bridge, but there was no bad weather in the forecast, so I ignored them. That may have been a mistake. A moment later, the ship rocks violently, nearly knocking me off the scaffolding's edge. Through one of the veranda's sliding glass doors, I watch as the swells grow larger. The weather has changed dramatically in just an hour. The beautiful sunset over the ocean has given way to dark, rolling clouds illuminated by strikes of lightning.

Having sailed boats my entire adult life, I know a fierce storm is on the way. Just as I think I'm getting the hang of the ship's motion, it tilts sharply again, and this time I lose my balance and fall through the scaffolding. Catching myself on the edge, I glance down at the ground far below, and I feel my heart start to race. Desperately, I try to pull myself up, but I slip again.

I hear Logan's voice over the radio. "Theo, we need you on the bridge right away. I repeat, Theo, we need you on the bridge now." Using every ounce of strength I have, I manage to pull myself up. However, the ship's rocking prevents me from maintaining my balance, and I have to crawl back to the elevator. Within seconds, I realize I'm trapped—the call button for the elevator is missing because the wall is still unfinished. Climbing back onto the scaffolding, I survey my surroundings for another way out. Looking down, I see the elevator on the 7th floor is complete and the call button is in place. Lying on my stomach, I swing my legs over the edge.

For the second time in just five minutes, I'm hanging from the scaffolding again. I hang there for a moment, trying to muster the courage to let go. I count down in my head, *three, two...* the radio blares again.

"Theo! Where are you?" Startled by Logan's voice, I let go, slamming onto the 7th floor, knocking the wind out of me and leaving me completely paralyzed for a few seconds. Struggling to catch my breath, I writhe around in panic. Just as I think I'm about to die, my body manages to take in a deep breath. I gasp for air, and the pain and panic begin to fade. Across the room, the radio lies cracked open with pieces strewn across the floor. I reach for the separated pieces and try to put the radio back together. Clicking the power button multiple times and hearing nothing, I let out a frustrated grunt. I take a breath and gently push the power button again. The radio powers up, and I press the transmitter button.

"I'm here! I'm on the 7th floor, Logan. I'll head your way as soon as I can. What's going on up there?"

Logan quickly replies, "Theo, are you okay? Can you hear me?" The radio's signal cuts in and out as Logan tries to speak. The signal completely drops. A few seconds later, there is static, then the signal comes back.

I quickly push the button to talk. "The radio signal is going in and out. I'm fine; I'll be up there in a minute. What's happening?"

The signal is stronger now. "A storm came out of nowhere," Logan says, his irritation loud and clear over the radio. "Whitaker is nowhere to be found, and I need everyone's help. Get up here as soon as you can." The signal goes out again.

The ship rocks violently, and I struggle to stay upright. Unable to stand, I continue to crawl across the floor. When I reach the elevator, I push the call button. Moments later, the elevator doors slide open and I crawl into the elevator, the doors slide shut behind me and I hear the cruise director's announcement over the intercom.

"Ladies and gentlemen, please seek shelter as we navigate through this weather. Ladies and gentlemen, please seek—" The announcement is cut off, and the lights go out. Complete darkness surrounds me. The elevator doesn't move. A few seconds later, the emergency lights flicker on, and the elevator doors slide open again.

The cruise director continues her announcement, "Everyone, please seek shelter." I crawl back out of the elevator and over to the wall blocking access to the 7th floor from the stairs and notice a small gap between two boards. I can fit my hands through and grip. Using all my strength, I pull as hard as I can, then suddenly there's a crack, and my entire body falls backward—the plywood breaks in half, creating a small hole that grants me access to the stairs. As I crawl through, a woman screams somewhere from the floors above, and the thought of Julie flashes through my mind.

I pull out my phone to text her. I type the message and press send. 'NO SERVICE' flashes in the corner, and the message fails to send. I notice the Wi-Fi is out, too. Frustrated, I throw my phone across the stairwell, where it shatters on impact. Angry with myself, I pick it up, see the cracked screen, and curse aloud. I need to reach Julie and Tyler somehow. Glancing at my watch, I relax slightly, knowing they've been together since dinner, which started twenty minutes ago. In the distance, I hear people screaming, and the sound of glass shattering fills my ears. I begin to panic again, my mind racing to the worst possible scenario... *The ship is sinking.*

7

JULIE

Sitting alone in the dining room, I gaze around, marveling at its beauty. Theo was right—he'd told me his father spared no expense making this ship extravagant. This dining room looks like it belongs in a fairytale. Positioned aft of the ship, it's bathed in the warm glow of the setting sun streaming through floor-to-ceiling windows, highlighting the ornate tiled floor. Each window has an intricate lattice pattern and thick navy curtains tied back with nautical ropes, enhancing the charm. The room features round and oval tables covered with white linens and faux candle centerpieces, creating a romantic atmosphere.

Sitting alone, lost in thought, I stare out at the vast ocean. I can see ominous clouds rolling in, but they make the sunset all the more dramatic, even if the weather causes the ship to rock a little harder. My tranquility is shattered when Tyler intentionally steps in front of the window, blocking my view.

"Hey," I say, looking at him, then at his sweatpants. "I see you dressed for the occasion." My voice was laced with sarcasm.

"Thanks, I'm comfortable," he responds.

"I'm pretty sure the dress code isn't sweatpants and t-shirts."

"It's okay. I know the owner, and I'm sure he won't mind," he says, winking.

"Look, Tyler, you may have handed everything over to your brother, but everyone on board knows you're a Parrish, so can you try and take things more seriously?" To avoid looking at him and getting mad all over again, I pick up my menu and study the appetizer list.

"Yeah, whatever," he says. I glance up and catch him rolling his eyes. "Where's Theo, and what are you going to get?"

Without looking up from the menu, I answer, "I haven't decided, and I'm not sure when Theo will be down. He should be here any minute."

"I feel awful. Do you have any Tums? I think I ate something that my body doesn't agree with."

"Nope. But I'm sure they'll have something in the medical center, or whatever."

"Okay," he says, his voice sounding a little weird. I look up and notice how hard he's gripping the back of one of the dining room chairs. "All this swaying is making me feel even worse. I'm going to stop there on my way back to my room." He closes his eyes and takes a deep breath. Come to think of it, he is looking a little green, I guess.

"Yeah, I hope the rest of the cruise isn't like this. I assume that's why Theo isn't here. He must be busy dealing with the weather, it–"

I'm startled by the sound of glass shattering behind me, and everyone turns to see what happened. A waiter is staring at the floor, an empty tray in his hands. His face turns beet-red, and he mumbles apologies as he squats down and starts picking up the shards of broken water glasses from the floor. Something catches my eye, and I look up to see the chandelier swaying. I frown, turning back to my table just in time to see the red wine in my glass shift and slosh up over the rim and onto the table. Just as I make eye contact with Tyler, my wine glass spills over, painting the white linen a deep maroon.

Throughout the dining room, the sound of glasses shattering alongside plates and cutlery crashing to the floor fills the air. "I've never seen a ship... tip this much." I take a deep breath, unable to stop my heart from racing. I stare down at the red wine stain blooming across the tablecloth like a bad omen. The entire room tilts hard to the right, and a woman at the table across from me screams as more plates fly off the tables. I ball my hands into fists to stop them from shaking. I really hope Theo gets here soon.

I look over at Tyler, he's gotta be as scared as I am, but he looks across the room, grabs an ice bucket meant for champagne and vomits, then sets the bucket down.

"Fuck. I feel better." He wipes his mouth on a linen napkin, "I'm sure it's nothing, just some big waves outside. The crew will probably steer clear of the storm at any moment. Try to stay calm."

I turn to my right and see a counter lined with metal water pitchers begin to slide violently to the left. I grip the table as tightly as I can, hoping it will somehow hold me in place. I feel my heart racing. I watch, as in slow motion, the ship tips again and the water pitchers slide to the opposite side, crashing onto the floor. People around the dining room stand and rush for the exits. Tables and chairs slide to one side of the room, some toppling over and crushing people in the process. Tyler, now panicked, grabs my arm and yanks me to the floor.

"You were right. Something's wrong. We need to find Theo. Now." His voice is trembling, and he has a frightened look I've never seen on him before. The look sends a chill straight down to my bones.

Quickly, I pull out my phone and tap Theo's name. The screen instantly flashes 'CALL FAILED' along with a low battery alert. I realize my service is out, but I try to send a text anyway. As soon as I hit the send button, my phone dies. Hoping it went through, I slide my phone back into my pocket.

"My phone died, but Theo should be upstairs at the bridge. Where's your phone?" I ask, doing my best to keep from freaking out.

"I left mine charging in the room. Come on, we can check the bridge. Let's take the stairs, the elevators aren't safe. Stay low," he says.

As we head toward the stairs, an announcement blares, instructing everyone to take cover. Suddenly, the voice cuts off as the lights flicker and the power goes out. Darkness engulfs the room, and the sound of screams and people running fills the air. In the chaos, a woman knocks me to the floor, and my left hand slams down onto a broken wine glass, the glass stem piercing through my palm like a nail shot from a nail gun. The pain is white-hot, and I scream. Moments later, emergency lights activate, illuminating everything. I see the glass stem embedded deep and scream again in shock. I grab the broken glass, biting my lip to hold back another scream. With a quick motion, I pull it out as fast as I can. Blood streams down my arm, splattering drops onto the floor. I look around and spot a white linen napkin on the floor and I grab it, wrapping my hand tightly. Wincing in pain, I clench the linen with my teeth and tie a knot to stop the bleeding.

"Are you okay?" Tyler asks.

"No, but we need to find Theo. I'll be fine," I say, trying my best to mask the pain as we crawl toward the dining room exit.

"We should head to the atrium and go from there. This stairwell's too packed with people trying to escape," says Tyler.

As I move cautiously down the hallway toward the atrium, I hear glass shatter again and more screams. For a moment, we stand, but as soon as we do, the ship lurches violently, knocking us into the wall next to us and back down to the floor. We are just a few feet from the atrium's entrance when the ship tips sharply again, and a loud crash reverberates through the hall. When we enter the atrium, I gasp. The elevators have torn free from their cables and are crushed on the floor. People are hysterical, running and screaming. Several individuals try to rescue two people trapped inside a crushed elevator; however, the shattered glass walls, still partly intact with large broken pieces, make access difficult. Some bystanders manage to reach inside and pull out one man. As he is lying on the ground, a woman begins CPR and calls for help. I'm paralyzed as memories come flooding back....

I'm twelve years old on a family road trip with my parents, a journey we'd been planning for months. My parents weren't wealthy, so they'd saved for over a year to make the adventure happen. It's our first family vacation in such a long time, and I'm beyond excited. The cross-country drive will take us all day and all night, and everyone called where we were going the most magical place on earth. I'm excited to finally be tall enough to ride all the attractions and can't wait to eat all the junk food. We'd been on the road for several hours and it was late at night, and I was hungry and cramped from the car ride. I kept whining for dinner—I was starving and cranky.

When my parents spotted a fast-food sign at the next exit, they told me I could choose any of them. We exited the highway and came to a stop at a red light. I spotted the glowing sign, and my stomach grumbled. I kept complaining about how long it took for the light to change, yelling "It's green, go!" once it finally did. My mom, behind the wheel, didn't check for oncoming traffic, and she hit the gas, and at the same time, another car sped through the red light and slammed into us.

In the backseat, I hung helplessly upside down, with all my weight against the seatbelt. Crying in pain, I reached for the buckle and pressed the release. Gravity pulled me down. Glass shards sliced into my arms and legs, and I screamed, feeling a sharp pain shoot through my stomach. I looked down to see blood oozing out with a large piece of glass protruding from my shirt. I looked up and saw both my parents laying motionless upside down in their seatbelts. I yelled their names in a panic to wake them.

My dad stirs, unbuckled his belt, and crawled out. He helped me out first and realized I'm injured. He stared at me silently, and I knew he was panicking. The pain finally stopped in my stomach when I stopped moving and the blood also stopped. I tell my dad that I'm okay and that I'll stay still until the ambulance arrives. I told my dad he needed to help my mother. Even though I'm terrified, I knew my mom needed more help. I couldn't see any blood on her, but something just felt wrong *when I looked at her. I watch as my dad runs to the driver's side but can't get inside because the car is crushed. He then runs back to the passenger side, shouting for help to the group of people who have gotten out of their cars staring at us. Shocked, scared, and trembling, I can't move. A few moments later, I watch my dad pull my mom out, lay her on the asphalt and start CPR.*

That memory has haunted my nightmares and kept me awake for years. As I got older, the nightmares gradually faded, but the memory never did. I've never told anyone that I remember what happened that night and most people only know that my mom was killed by a drunk driver. If anyone asked me if I recalled it, I told them that doctors say my brain might have blocked it out.

My dad never fully recovered from the loss of my mom and later became an alcoholic. I basically raised myself from age twelve and haven't spoken to my father in over a decade. I vividly remember every single detail, from the moment the car hit to the ambulance's arrival and the paramedics telling my dad that his wife didn't make it. I remember all of it...

An ear-piercing scream in the distance brings me back to the present, and I watch as another group of people tries to pull the other person out of the crushed elevator. Two bystanders grab the trapped woman by her arms and begin to pull. I watch as they pull the semi-conscious woman out, and her torso becomes caught on a thick shard of glass. The people pulling must not realize this, so they continue to tug. Before I can shout for them to stop, they pull even harder. Her torso is sliced open, and blood spurts out of her chest. The woman suddenly lets out a blood-curdling scream. I can't take my eyes away from her, and within seconds, she loses so much blood that her face turns a ghostly white, and she stops moving. Her body is bleeding out onto the atrium floor. The wound is massive, and her intestines are visible. Her body lies motionless, slumped over the elevator. A man from across the room runs up and drops to the ground beside her, sobbing. As I stare at the man, it crosses my mind that he might be her husband.

Then I wonder where the hell my husband is in all this madness.

8

THEO

Think, Theo, think! Julie thinks I'm on the bridge, and without any way to communicate, that's where she and Tyler will go to look for me. That's where I need to go. While climbing the stairs, the ship rocks violently, and a woman trips a few steps above me, tumbling down the stairs and nearly taking me with her. When I look back at the bottom of the stairs, I see she's unconscious. I change direction to head towards her. Hovering above, I gently shake her shoulder.

"Ma'am, are you okay?" When she doesn't answer, I try again. "Ma'am, are you okay?" Still no reply. I check her pulse, but feel nothing. *Oh, shit.* As seconds pass, more passengers start to flood the stairwell. I grab the arm of one of them, asking them to help the woman. "I have to go, I need to get to the upper deck," I say, as I head off in that direction, with screams echoing behind me as the person I stopped probably just realized the woman is dead. Maybe I can fix this. But how? I only have a few steps to reach the bridge door. *Stay calm, Theo. Stay focused.* Almost there, I grab the railing and lift myself toward the bridge door. I swipe my key card—nothing happens. Frustrated, I pound on the door with my fist.

"Let me in, it's Theo." The door swings open, revealing utter chaos.

Logan's eyes lock onto mine as soon as I step inside.

"We've lost all engine power," he says, bracing himself against the navigation panel. "Whitaker put me in charge until he gets back. He popped back in to say he's heading down to reset the breakers and check on what's going on in the engine room. It happened when the lightning hit the ship, causing that huge power failure." As I scan the room, taking in the damage, I notice the furniture has been knocked over and scattered to one side, in complete disarray. Bolting down the furniture was supposed to be done before we sailed, but with our accelerated timeline, we skipped it for this practice run and only bolted down the captain's seat.

The glass windshield on the bridge has a huge crack that's easily three feet long, likely caused by the impact of a piece of furniture. Two officers who were working with Logan are now lying on the floor. One officer is unconscious, blood dripping from the side of his head. I can't tell who it is, but I think it's Greg. The other officer is screaming in pain–a broken bone protruding from his right leg. Logan quickly rushes over to help the officers on the ground. My heart is still racing, I feel the room spinning, and I become lightheaded. I try to steady myself, grabbing onto a chair that's still upright, while Logan screams in the background. Standing there, my heart pounding, I close my eyes for a moment and take a few deep breaths, then I repeat the process—something a therapist taught me for when I get overwhelmed.

With as much calm as I can muster, I turn and ask Logan, "What can I do to help? Where do you need me?"

Logan points to the radio by the control panel. "Get on the radio and put out a mayday call for help." I sprint toward the panel, but before I reach the radio, a bright orange glow erupts outside, followed by a loud bang. A massive fireball explodes, sending me tumbling to the ground. The bridge's glass shatters inward, pelting everyone with shards. I look around, realizing my vision has narrowed to a tunnel. My head throbs, and the taste of iron fills my mouth. I reach up and feel my face. Blood drips from my nose, and my upper lip is split open. My ears are ringing, and I can see people's mouths moving, but I can't hear anything.

I glance over and see Logan screaming in my direction, and I blink in confusion. A few seconds later, the ringing subsides, and Logan's shouting fills the air, asking if I'm okay. Outside the destroyed window of the bridge, people are screaming, running, and begging for help. Smoke fills the bridge's interior, making it nearly impossible for me to breathe. My instincts kick in, and I keep low, reminding myself to *stay calm*. I crawl toward the door and kick it open. Despite the pitch-black night sky, the fire on the top deck burns so brightly that it lights up the ship's 12th deck like the heart of Times Square. The officer who was unconscious on the floor earlier stirs and looks around.

Before I crawl outside, I shout to the officer, "Greg, grab the sat phone next to you and call for help." When I look back at Logan, I see him lying on the ground, a large piece of glass protruding from his right thigh. I wince at the sight, suddenly feeling queasy. Logan lets out a piercing scream as he yanks the four-inch shard of glass from his thigh. I watch as he tears off his pant leg and ties it around his wound, creating a tourniquet. Then, as if barely injured, he begins crawling toward me.

I hear screaming behind me, and I turn around to see a man running toward me, begging for help. His face is covered in blood, and his left ear is torn off. Still on the ground, I quickly crawl backward, stunned by the gruesome scene. I then quickly stand up and push past the man to get out. I grab a pool towel lying on the ground and use a broken piece of glass to cut a strip of the towel off. I run back to the man and make a bandage to help protect his bleeding ear.

Standing outside the wrecked bridge, I have a clear view of the entire deck. The sight of the destruction leaves me speechless. The wind is so fierce it almost knocks me over, and the rain pours down so heavily that each drop feels like needles piercing my skin. The ship continues to sway wildly, rocking side to side, out of control. A steady river of rainwater mixed with blood flows across the deck, and people are lying all over the deck in every direction, with blood streaming from their bodies. I watch a man on fire run across the deck. It's like it's happening in slow motion, like I'm watching a movie I can't stop. The burning man drops to the ground near me, screaming in pure agony, and begins to roll around on the wet deck, but the water is no match for his clothing engulfed in flames.

Suddenly, I remember there's a fire extinguisher behind every bar. I run as fast as I can, jumping over bodies, to the pool bar in search of the extinguisher. Looking around, there are none in sight. *What the fuck?* I look back at the man, who has now stopped screaming and moving. He's lying on the deck, face up, burned. The rain extinguishes the fire, but it's too late. Unable to look at him any longer, I turn to the pool and notice that the waterslide is also completely destroyed; the explosion blew it apart into several pieces strewn across the deck.

A large piece of tubing has a screaming man pinned to the ground. I abandon my search for a fire extinguisher and rush to the man's rescue. I try to lift the tubing, but it's too heavy. I look around, shouting for help, but several people are running past me. I grab a person nearest to me, a guy who looks like he's in his mid-thirties. He's not bleeding or anything, so I hope he's okay. "I need your help. Please!" The man nods and runs around to the other side of the tubing. We push and shove at the tubing, all the while the screams of the trapped man keep drifting up around us. A few others notice as we struggle to lift the heavy, broken tubing and finally join in.

I shout, "Lift on the count of three. One... two... three, lift!" We all lift together, and the slide rises a few inches, just enough to pull the man out. As he's being pulled free, the onlookers gasp at the sight. Looking down, I realize why. The man hasn't seen his body yet. When he looks down, he finally sees his injuries, and he screams again before passing out. His left leg is completely crushed, with bones and muscles exposed, and his right foot has been severed. *Oh my god.* My stomach roils, and before I can stop myself, I lean over and vomit onto the deck. I wipe my mouth with the back of my hand, overwhelmed for a moment by the chaos around me. People are running in every direction, frantic with fear. The ground is littered with dead bodies, and the air is filled with agonizing screams and desperate pleas for help. I try to focus, to find something to concentrate on. *Who can I help?*

To my right, I hear a woman's scream cut through the noise. I turn and watch as panicked people trample her. My feet are moving before I can think, but by the time I reach her, she's stopped screaming and is no longer moving. I shove the people away, praying she's still alive. I see her face is badly battered, and her arm is twisted unnaturally behind her back. I reach down and place two fingers on her neck, checking for a pulse. Fuck. She's dead.

My attention is drawn when I hear someone call my name. "Theo! Over here!" I scan the deck and spot Greg—the officer from the bridge who grabbed the satellite phone.

He is shouting at me, holding the satellite phone high above his head. "I can't get it to work."

I wipe the rain from my eyes as I think about why it might not be working. Damn, what did they say in that training we had? Oh, right! It needs an unobstructed signal. I notice that Greg's standing under an awning. He needs to be in the open.

I shout back at him, "Get out in the open!" Greg rushes to the edge of the deck, lifting the phone and staring at the screen. I run toward him and see the device's red light flash to green, indicating a stable link.

"I've got it! I've got it!" Greg waves the phone at me.

I watch as he fumbles with the slick phone in the rain. It slips from his grasp and hits the floor. He quickly picks it back up, but the green light turns red again —signal lost. Greg moves closer to the edge, holding the phone high, trying to establish another signal. While he is trying, I watch as a group of panicked passengers rushes past him.

"Wait...watch out, Greg!" He turns to face me, and someone staggers into him, knocking the satellite phone overboard.

I fall to my knees, utterly stunned. People are dying all around me, but I can't think straight. I don't know what to do. Lightning flashes like a strobe, thunder rumbles, and people scream. Rain pelts my head and streams down my face. A loud banging sound startles me. I look around, but there's no sign of anything that would have caused the sound.

BANG!

The sound reverberates again, and this time I hear it coming from the side of the ship. I hurry to the balcony and look down, seeing hundreds of people struggling in the water. Multiple passengers wear bright orange life jackets with flashing lights, while others have nothing. They all struggle to stay afloat in the rough waves.

Passengers wearing life jackets are being attacked and pulled underwater by others trying to steal them.

BANG!

What the fuck *is* that? I look around again and notice a lifeboat swinging back and forth, crashing against the side of the ship. The rough seas make the ship tilt repeatedly, causing the hanging lifeboat to sway.

BANG!

A large section of railing breaks off the 4th floor, slamming into the ocean. All the other lifeboats are being lowered. More railing breaks off, sending more passengers overboard. As another lifeboat is lowered, its securing wires snap, and it plummets into the ocean, crashing down on several passengers below. Most of the lifeboats have been lowered improperly, causing them to fall several stories into the water. The impact shatters their fiberglass shells, causing them to take on water and sink.

I watch as one lifeboat crashes into the water. The moment it contacts the ocean, chaos breaks out in an instant, water splashing everywhere. Fights erupt on the lifeboat as people scramble to escape the rough waters. It looks like a pack of piranhas attacking their prey. Out of the corner of my eye, I catch a glimpse of something glowing. Looking over, I see a flare flying through the air, its red glow lighting up the dark sky before plummeting into the ocean seconds later. The flames from the explosion are dying down as rain pours onto the ship. I look around, searching for the blast site, and see one of the smokestacks, with metal rods protruding like needles in a pincushion. Where the hell did those rods come from? I look back at the bridge, completely destroyed. Next to it is the deck that houses the ship's communication gear and radar systems.

The massive white sphere housing the radar system is missing. The antennas are bent and probably useless now. The steel railing surrounding the equipment is also destroyed, with missing sections. I look back and forth between the smokestack and the destroyed upper deck and...holy shit. Those aren't rods, they are deck rails. The explosion must have happened on the upper deck beside the bridge, which explains why the impact was so intense inside the bridge. My mind races with questions. *How the hell could this happen?*

A bolt of lightning illuminates the sky followed by the sound of popping that startles me, bringing me to my knees. Turning towards the noise, I notice the lights around the pool have all blown. Standing up, I look back at the rails pierced into the smokestack and realize everyone will likely assume the explosion was caused by lightning striking the ship. Even after witnessing a strike hit the ship I know it's nearly impossible for lightning to cause such a massive blast in that particular area. I run through every possible scenario that could explain the explosion. The only thing that makes sense is that the explosion wasn't an accident—it was intentional. Someone must have set off a... bomb beside the bridge, maybe? To take out whoever was inside or to take out all communications. But why? Thinking about that will have to wait until things calm down.

Scanning the deck, I assess what's missing: the lifeboats have either been deployed or damaged beyond repair, and the radar and communication equipment have been destroyed. The ship has lost all engine power; the electricity is out and only the generators are still working. Without power, the internal radios can't connect, and there's no cell service. This means we're cut off from communicating with anyone on the ship or on land.

As I consider this, my family flashes in my mind, and I decide my top priority is to find Julie and Tyler, then develop a plan to survive. I take a deep breath and start looking more critically at things. The ship doesn't seem to be sinking. The explosion happened on the top deck, not below, so likely there's no major damage to the hull of the ship. At least not enough that it'll sink right away. This is good, but how do I get everyone to calm down? *Think...think, Theo.*

Staring at the bridge, I recall the checklist Logan had me run through. One by one, I think through each item. "The megaphone," I say out loud as soon as the thought enters my mind. I start running toward the bridge, hoping the equipment cabinet hasn't been damaged. I'm halfway there when, out of nowhere, someone grabs my arm and yanks me backward.

9

JULIE

As the ship continues to tilt violently, furniture from the upper floors crashes down into the atrium. Everyone rushes to find cover. I look up just in time to see a solid white marble table flip over the 6th floor balcony and smash into the frosted glass stairwell, shattering at least ten steps. Those standing on the stairs are knocked down, and the glass shards slice several people as they descend. I watch, horrified, as crimson blood sprays, painting the frosted glass. The ship tilts again in the opposite direction, sending more furniture crashing down and crushing passengers along the way. Another massive marble table hits the stairs, this time smashing into the railing. The wooden railing splinters, leaving one side jutting toward the ceiling like a spike.

Behind me, Tyler vomits again as he watches the chaos unfold. He wipes his mouth, points toward the front of the ship, and shouts for us to keep moving. To my right, round portholes line the hallway. Outside, the rain pelts the window with such force that it looks like a fire hose spraying. A moment later, the night sky illuminates, causing the hallway to suddenly glow bright orange, followed by a loud explosion.

"What the hell was that?" Tyler asks, his voice shaking and

with fear in his eyes. I stay silent, my gaze fixed on the stairs at the end of the hallway as I crawl forward. My body trembles with fear. Out of the corner of my eye, I catch a glimpse of movement. As I turn to my right, I gasp, then cover my mouth.

"What's wrong?" Tyler asks, his eyes still fixed on me.

"Look out the window," I whisper.

Unable to look away, I see hundreds of people struggling to lower a lifeboat. Amid the chaos, the cables break free, and the lifeboat crashes into the railing, crushing a group of passengers below. Blood splatters across the face of an elderly woman standing nearby, and she lets out a piercing scream, almost loud enough to shatter glass. The sound is so intense that I hear it through the window as if I were standing beside her. Behind her, several other people grasp the broken railing, desperately trying to pull themselves up to prevent falling to their deaths.

As the ship rocks violently, the lifeboat swings back and forth off the edge, destroying anything in its path and pushing passengers overboard into the water.

"We've got to keep moving, the stairwell isn't far. Stay close to the wall so you don't get trampled," Tyler yells in a commanding voice. I follow his instructions, keeping my head down and my shoulder pressed against the wall to avoid putting pressure on my injured left hand. I crawl until I find an opening to the stairwell. As we enter, we navigate around people who have fallen or been trampled, their lifeless bodies scattered all around us.

Most have twisted limbs and all are drenched in blood. I glance back to check that Tyler is still with me, only to notice my own dress and legs are stained with blood. Suddenly I feel my heart start to race and I begin breathing rapidly, I roll over and pull up my dress, rubbing at the blood on my legs frantically, searching for the source.

Tyler grips me tightly, holding my shoulders to get my attention.

"Julie, you're okay! Breathe with me." Our eyes meet and I take a deep breath. I look down at my hands, now covered with blood.

"I don't know whose blood this is. Is it mine?" I ask, my voice nearly cracking and my hands trembling.

"No, look around; people are bleeding all over. Does anything hurt?" he asks, scanning me for signs of injury.

"No, I think I'm okay. My hand hurts, but that's about it," I say, trying to calm down. My racing heart starts to slow, giving me a brief chance to catch my breath.

"We need to keep moving. People are still running, so be careful and watch out," Tyler says, lifting me to my feet.

"Thanks, Tyler," I reply softly.

"You're family, Julie, and we've got to look out for each other."

I give his arm a quick squeeze. "Come on, let's go find Theo and figure out what's going on. Are you alright?" I ask, realizing I haven't checked in on him since the chaos began.

"Yeah, I'm fine."

Following Tyler, I watch as he shoves an unconscious man out of our way on the stairs. The man's body sprawls across multiple steps, his feet higher than his head. When Tyler moves him, the man rolls over. When his face comes into view, I'm puzzled for a moment by the man's familiarity. *Where have I seen him before?* I stop climbing and look at the man, trying my hardest to place where I know him from.

"Let's go!" Tyler shouts, jolting me out of my daze. As I keep climbing and accidentally bump into the familiar man, a white key card falls from his pocket. I pick it up and realize it's a crew ID. It's the bartender who'd served us drinks earlier. Damn.

"Let's go, Julie," Tyler shouts again.

"Coming, I'm coming."

The ship has calmed somewhat, and we are able to stay standing for a moment.

"Take advantage of the calm while we can," I say loudly. Together, we quickly make our way up the stairs. One by one, we dodge bodies sprawled on the floor and passengers rushing in the opposite direction. I'm stopped when a woman grabs me, screaming for help. I'm shocked to see that she's missing some of her fingers and has burns all over her body and half her face.

I cringe at the sight and try to stop and help. Tyler grabs my arm, not letting me. "We don't have time, Julie. We need to find Theo first. We can help others afterward."

A few steps later, another woman is screaming for Tyler's and my help. Her leg is bent unnaturally backward at the knee, broken in multiple places.

"Please help me, please," the woman begs, sobbing. I look back at her as Tyler pulls me away. I open my mouth to tell the woman I'll come back and help, but in that instant, I realize I'm not sure I *can* come back and help. Suddenly, the gravity of the situation hits me like a freight train. I stumble to a halt, taking a look around. I'm surrounded by blood and bodies, with people screaming for help in every direction.

"Is the ship sinking?" I say, my heart racing as I try to take it all in. "We need life jackets! Where are they?" *Come on, Julie. Think. What was in that safety drill...* "They're in the rooms! We've got to get to the rooms."

"Julie, we don't have time! We've gotta find Theo, come on!" Tyler yells in my face, so close I can feel his hot breath. I look around frantically, trying to find anything to grab on to.

"There," I say, pointing toward two bodies lying motionless in the hallway. "Let's take those from them."

Tyler spots what I'm pointing at and nods. He rushes down

the stairs, darting across the hallway toward the bodies. At the same time, I see a man quickly approaching. He must have the same idea. I watch as the burly man tackles Tyler to the ground, stopping him in his tracks. With no time to lose, I race to the bodies. Once I reach them, I turn back to see Tyler on the ground, grabbing the man's foot to keep him from coming toward me. The burly man fights back fiercely, trying to escape Tyler's grip. The man looks back at Tyler and kicks him in the face. Tyler shouts in pain, grabbing his nose. Blood starts pouring down his face, soaking his chin and the collar of his shirt.

After the man stands up, he sprints straight toward me with a vengeance. I struggle to unclip the vest from one of the bodies, trying and failing. Out of the corner of my eye, I spot a champagne bottle rolling back and forth with the ship's tilt. I lunge for the bottle as the man closes in on me. I grab the champagne bottle by the neck, realizing it's full and has a bit of weight to it. Without looking, I spin around, wielding the bottle like a baseball bat. The bottle shatters against the side of his head, slicing his face in the process. The man drops to the ground, unconscious. I watch his chest rise and fall for a moment before I look back at Tyler, who's running toward me.

"Sorry, I tried to stop him," he says.

"It's fine. I'm having trouble with these damn clips; it's hard to do with one hand. Go get the other vest?" I say, frustrated, fidgeting with the clip.

Tyler crawls over to the other body, then looks over, grinning at me. "That was pretty badass what you did with that bottle, Julie."

"Thanks, my fight-or-flight kicked in."

"I'd say the fight response kicked in, not the flight," he says just as the vest in his hands unclips.

"What's wrong with this thing? Ugh!" I shake the clip, getting even more frustrated.

"Here, let me help," he says, pressing firmly on the button, and the clip separates. We put on the life vests and buckle the clips back on.

At the top of the stairs, people rush inside from the 12th deck. Tyler and I quickly press ourselves against the wall to avoid being knocked over in the chaos. The smell of smoke fills my nose, and heat washes over my skin from the flames flickering. I catch a glimpse of the 12th deck and see the bright red inferno raging across the entire deck. Tyler and I make our way out the door into complete and utter chaos. I step out onto the fiery deck and I'm instantly overwhelmed by screams, even louder than inside the atrium. My eyes sting from the smoke billowing all around me. My vision blurs from the heavy rain and wind whipping me. The wind is so fierce that I feel my feet slipping on the wet ground as I struggle to stay upright. I hear Tyler shouting behind me, but I can't make out what he's saying.

I take in my surroundings. The water slide lies in ruins, scattered across the deck. Tables and lounge chairs are overturned against the walls and railing, battered by the wind. Dead bodies are everywhere, in much worse condition than those in the atrium or hallways. I look down at my hands and watch, mesmerized, as rainwater washes the blood off. My makeshift bandage, once white, now a light pink, reminds me of a watercolor painting. My wet hair hangs in front of my face, clinging to my cheeks. The taste of iron fills my mouth. The thought of someone else's blood dripping into my mouth crosses my mind, and I instantly feel my stomach churn. To my left lies a woman, face up, with one hand missing. My eyes continue down her body, and I realize that half of her leg is also missing. The sight of the severed body parts makes me gag.

I feel bile rise in my throat, making me look away. Before my brain processes what's happening, a flash of white light appears miles above me, followed by a loud clap of thunder. An instant later, another lightning bolt strikes, but this time it's not above me—it's directly in front of me. A pole with a string of lights attached brightens with a surge of electricity, then, one by one, I watch each bulb pop loudly, sending sparks flying. Instantly, I drop to the ground, pulling Tyler down with me. My knee lands on the dead woman's torso. I quickly scramble off the woman and shout to Tyler, asking if he's seen Theo. Tyler remains silent as he lifts his hands from the slippery muscle fibers hanging from the woman's leg. In an instant, Tyler crawls away. Before he stops, he projectile-vomits a third time, splattering the ground.

He stands, wipes his mouth, and looks at me, his face ghostly white. "Sorry, I just couldn't hold it back anymore."

I pat him on the back, unsure what else to do. "It's okay. I don't think my brain is even processing what's happening. Like, I'm on autopilot." I look past Tyler and see the remains of the bridge. "I think I see Logan," I shout.

"Who the hell is Logan?" Tyler asks.

"He works with Theo," I say, running over. When he comes into view, I notice one of his pant legs is missing, and his white suit is now pale pink from the rain and blood. He's limping, with a cloth tightly wrapped around his right thigh above the knee, a dark red stain seeping through.

I look up, ignoring the wound for the moment. "Have you seen Theo?"

Logan shakes his head. "No, I haven't. We were on the bridge when the explosion hit. Everything happened so fast. People were hurt and screaming. I was knocked down, and then I couldn't see anything."

"Do you have any idea what's going on? How did this happen?" I ask, assuming he's the most likely to know.

"Honestly, I have no idea," he says, a confused look crossing his face.

Tyler chimes in, "Where's my Uncle Mark? I mean, Captain Whittaker?" We both stare at Logan, waiting for his answer.

"The last time I saw him was just as the storm hit. Lightning struck the ship and knocked out all the power. He said he was going to reset the breakers that control the engine's power and left me in charge until he got back." Logan shifts his leg, wincing in pain. "Right after that, everything went crazy. He could be anywhere." I wipe away the hair sticking to my face.

"We need to find Theo first, and then we'll worry about Captain Whittaker. Can you keep going? Your leg looks pretty bad."

"Yeah, I'm fine. I can walk."

"We should stick together to find Theo and Mark," Tyler says.

"That's a good idea," I say.

A loud bang echoes every few seconds from the side of the ship. The sound brings back memories of the lifeboats crashing into the hull. As we survey the scene, the fire begins to dwindle in the rain. Bodies float in the pool, and people cling to lifeless loved ones scattered across the deck. In the distance, another brawl erupts over a life vest on a crew member's corpse. I watch as two men grip the vest, pulling so hard that it rips in half. They let go and rush in opposite directions.

"Theo!" I hear Tyler shout. I look over at him, following his gaze, and see my husband staring at the bridge.

"Theo! Over here!" I shout, my voice straining and cracking. He must not hear me. Tyler and I keep screaming his name, but nothing happens. Then I watch him take off like a bullet toward the bridge. Still shouting his name, I chase after him. When I finally reach him, I grab his arm and pull him to me. He jerks toward me, slamming into my body as we both slump to the floor.

10

THEO

Lying on the deck, I face the sky as rain pelts me in the face. For a split second, I want to lie there and just give up. Then my thoughts shift, and I want to know who grabbed me. I quickly turn and, to my surprise, find Julie lying on the ground next to me. We both stumble to regain our balance in the pouring rain. As soon as I'm upright, I hug her as best I can around the bulky life jacket, and we stand there for a moment. "Thank goodness you're okay," I say, squeezing her. Then a thought strikes me. "You need to take this off immediately," I demand, grabbing at the life jacket.

"No," she says sharply, "I literally had to pull this off a corpse while fighting a man for it in the process."

"That's exactly why you need to take it off. People are getting hurt over these things. The ship isn't sinking, so you don't need it."

"Wait, how do you know the ship isn't sinking?" Julie asks, then quickly adds, "Are you okay?"

"Yes, I'm fine. I don't think I'm hurt, except I'm having a hard time hearing right now," I answer.

"Okay, I think that's good. So how do you know that the ship isn't sinking?" she asks.

"Because the explosion came from beside the bridge and destroyed everything up here. There shouldn't be any damage that would cause the ship to take on water."

"What caused the explosion?" Julie asks, wiping her hair out of her face and taking the life jacket off.

I pause, considering how to answer. I don't know if I should share my theory or keep it to myself. If I tell her, it might scare her. "I'm not sure." That's technically the truth. I *don't* know what actually caused the explosion. "Where's Tyler, and what happened to your hand?" I ask, changing the subject.

"When all the commotion started, I fell and my hand slammed down on a broken wine glass."

"Are you okay? We can try to find a first-aid kit and find something to clean it with, so it doesn't get infected."

"I'm fine, we've got bigger problems right now than my hand. Tyler is over with Logan," she says, pointing towards Tyler.

"Okay, wait right here," I tell her as I turn toward the bridge.

"Where are you going?"

"I need to grab the megaphone from the equipment cabinet. I need a way to get everyone's attention. I'll meet you with Tyler and Logan in a second. I love you."

"I love you too," she responds. I give her a quick kiss then run toward the bridge.

The bridge looks even more damaged than before the explosion. Just inside the door, another officer is lying on the floor, not moving. Blood has pooled around his head. I bend down to check his pulse, even though it's clear he's dead. With time running out, I look around and notice the emergency equipment cabinet has been opened and most supplies are gone. The only thing left is first-aid kits. My eyes drift to where the survival rafts should be before spotting the red megaphone in the top right corner, strapped to the wall. I reach up and unstrap it, then quickly press the button to test it.

"Testing," I say into the megaphone. My voice echoes throughout the room, and I wince from the volume. My left ear is sensitive to sound, while I still can't hear anything in my right. I take a moment to gather all the available medical supplies and first-aid kits. Surprisingly, there are several kits left. I grab a nearby trash can and toss the contents. I throw all the medical supplies and first-aid kits into the bag, lift it over my shoulder, and rush back outside. The chaos has subsided somewhat, and the fire is nearly out due to the rain. Only a little smoke and a small fire remain. Most of the passengers have fled the deck. The only people still screaming for help are the severely injured or those clinging to someone for support. I try not to look at the dead bodies on the deck. Holding the megaphone above my head, I press the siren button, letting it blare for a few seconds. The siren grabs everyone's attention on the deck, just as I wanted.

"If you can hear this, please make your way inside and head to the atrium. I repeat, if you can hear me, please go to the atrium," I say, hoping my voice reaches everyone. I watch people start to walk toward the exit. Most of the passengers are covered in blood, with cuts and bruises. Some have serious injuries, like broken arms, legs, or severe burns from the explosion. A few are even missing limbs and are being carried by someone else.

"If you see someone injured, please help them to the atrium," I repeat until everyone is helping someone. Looking around, I spot Julie, Tyler, and Logan standing to the side. I walk over to them and ask if they can help gather people and lead them down to the atrium. They quickly tell me what a disaster the atrium is, but it's too late to do anything about it—people have already started heading that way.

I continue with the megaphone, "If anyone has medical training, please form a group to the side." To my surprise, a few passengers do, and one woman approaches me and introduces herself.

"Hi, my name is Mary. I'm a retired nurse. How can I help?"

I watch her speak; her authoritative tone is like that of a teacher. The crow's feet and wrinkles around her eyes show she smiles often, and her smooth mahogany-toned skin gives her a youthful look. Her hands and clothes are splattered with blood, clearly indicating she has already helped multiple people.

I hand her the trash bag filled with medical supplies. "This is all I can find for now. We can head to the medical center for more supplies. There is a whole slew of hospital-grade medical supplies there. Can you make this work until then?"

"I'll do what I can," she says with a sigh, looking around at the carnage. "Some of these people need a hospital to survive."

I run my hand through my hair and let out a breath. "Well, ma'am, we don't have a hospital, just the supplies in the medical center. Please do what you can with this and take the supplies to the atrium. Thank you for offering to help," I say.

She reaches out, laying a hand on my shoulder, and her touch is soothing. "Take a deep breath, honey. I'll do what I can to help." Then she grabs the bag and takes it with her.

Before she leaves, I put my hand out, stopping her. "I appreciate whatever you can do."

Standing on the deck, I look out over the water. A few

lifeboats drift away from the damaged cruise ship, slowly disappearing into the darkness. As I walk toward the railing, I think back to earlier, when I watched passengers desperately trying to climb into one of the lifeboats. Gazing over the railing, I'm struck by the sight of bodies—hundreds thrashing in the water, while passengers around them scream for help, their life vest lights blinking brightly. I feel the wind suddenly pick up and watch the rough sea pull them farther away. The heavy rain makes it nearly impossible to see exactly how many are alive, and the flashing lights on the life vests make it even more difficult to keep track. They flicker so sporadically that I lose count. The current slowly carries the small, flashing lights farther out, and as I watch everyone in the water, the scene reminds me of an airport runway, with its lights flashing across the ground.

While the screams slowly fade into the distance, I close my eyes and remind myself to focus on the passengers still on the ship. I look down and realize I can see the 4th floor deck. The rails are damaged, and several lifeboats hang precariously. I spot a few passengers and try to catch their attention. With the strong wind and heavy rain, I'm not sure if they hear me. My voice strains from yelling, and I completely forget I'm holding a megaphone.

I yell into the megaphone, "Hey!" A few passengers look up in my direction.

"Head to the atrium and tell everyone you see to do the same." I watch as the passengers below scramble like ants. I turn to face Julie, Tyler, and Logan, who are now just feet away, trying to help anyone around. "Can you all walk to the atrium?" I ask.

"Yeah, we're all good. I might be a little slower, but I'll manage," Logan says, grabbing his thigh.

"I'll help you if you need it," says Tyler, offering a hand.

"No, I'm good, but thanks. Go help someone who needs it. I'll be fine. It's just sore," says Logan.

As we head to the atrium, we enter the side stairwell, partially lit by emergency lights that will go out soon. Like the deck, the stairwell is littered with bodies in various positions. One by one, we stop to check each body for a pulse. Only a few passengers have one, and we help as many of them as we can to the atrium.

As we enter the atrium, I realize Julie and Tyler weren't exaggerating about the destruction. People are huddled in small groups off to the sides wherever they can find space to stand or sit. The remnants of the two elevators and their shafts make the section of the atrium near the stairs almost impossible to cross. I watch as people hug the walls to navigate to other areas, trying to avoid the grand staircase, which now looks like something out of a slasher film. Although there are fewer dead bodies in the atrium than on the 12th deck, since we moved everyone to this area, almost every surface is stained with blood. I stand frozen, taking in the scene and feeling completely overwhelmed. I'm not sure how much time passes before my brother snaps me back into reality.

"You're the one with the megaphone. What do you want to do?" asks Tyler.

I look at my brother as Julie and Logan walk up and stand beside him. Both of them are silent as they wait for me to answer Tyler's question.

"We have to make this area work. Does anyone know where Captain Whitaker is?"

The three of them look at me and shrug.

"Okay, Tyler, you're coming with me to find him. Julie, can you start getting the passengers who can move to clean up the place as much as possible?"

"Of course, I'll do what I can. Theo, please be careful," she says.

"I will. I think we've made it through the worst of the storm, and I think it's a good idea to gather everyone in one place to make a plan," I say, looking around at all the people standing around. "Logan, do you think you can help the nurses set up a medical station?" I glanced at his leg. "Afterward, you should get yourself checked out."

"Yeah, I can get that done," Logan answers, then turns and hobbles away.

I glance at Tyler and motion for him to follow. We make our way toward the stairs and descend three floors. With the power out, the security system automatically unlocked all the doors. When we reach the 1st floor, we walk to the door marked 'Engine Room, Crew Members Only.' The door is sealed with a quick-acting watertight hatch that has a manual twist handle. Tyler and I twist it together, and the heavy steel door opens outward. A strong smell of something burning, mixed with the scent of burnt rubber, fills my nostrils. I try my best not to gag.

"What's that awful smell?" Tyler says.

I quickly respond, doing my best to keep my nose and mouth covered. "I'm not sure. I know the rubber scent, but that other smell is awful."

I can't see anything; the emergency lights aren't working. I reach for my belt, remembering I have a flashlight, and pull it out. I shine the light around the room, and I'm speechless for a moment. The first thing that comes to mind is the powerful lightning strike that hit the ship earlier.

"The lightning strike must have sent a surge so powerful it melted all the wires, and that's the burnt rubber we're smelling."

Tyler gags. "Can we just find Uncle Mark and get out of here? I'm not even sure if there is anything left for me to vomit, but I'd rather not find out. I don't think I'll ever get this smell out of my nose."

I shine my flashlight all around as we walk, my footsteps echoing through the vast engine room. The space is filled with compartments that control the propellers, and the electrical circuit board sits nearby. As I shine the light down the long corridor, I spot the breaker panel at the opposite end of the room. The panel door swings back and forth with the ship's motion, letting out an eerie, prolonged squeal. The sound is so creepy that the hair on my arms stands up.

"Are you still with me?" I ask, making sure Tyler is still next to me since he isn't speaking.

"Yeah, I'm right here," he answers.

As we approach the breaker panel, the rancid smell intensifies with each step.

"I'm not sure how much more of this I can handle," says Tyler, still gagging. If he doesn't stop, I'm going to throw up just from the sound.

"We're almost at the breaker panel; let's check if any switches are blown. I think the lightning damaged everything, and I don't see Mark," I say, looking around. We finally reach the panel, and I shine the light on it. Just as I suspected—every fuse is blown. "Can you help me hold the light, or come over and flip a few breakers to see if anything happens?"

"Yeah, I can do that," Tyler says between gasps, holding his breath. As he walks toward me, he accidentally kicks something on the ground. It flies across the room and hits the wall with a loud thud.

"What was that?" I ask.

"No idea," he answers, sounding as confused as I am. "I can't see anything down here. Bring your light closer."

I walk toward him, roughly ten feet away, and illuminate

the ground where the sound originated. As I scan the area, I suddenly freeze when my light lands on the object.

"Is that his shoe?" Tyler asks as soon as he can see what has my attention.

"I think it is."

"Why would he leave with only one shoe?" Tyler asks, his voice soft, confused.

I turn away, and my light falls on the scene in the distance. I can't bear it anymore, feeling my stomach churn as I realize the source of the terrible stench. Suddenly, I vomit uncontrollably.

"You okay, dude?" Tyler asks from behind me. I stand back up, wiping my mouth and watery eyes.

"I don't think he did," I whisper.

"What? What are you talking about?" Tyler walks up behind me. Before he gets to me, I try to stop him from getting any closer.

"No, Tyler. Don't..."

Tyler snatches the light from my hand and shines it where I was just looking. I watch as his shoulders drop the second he sees what I saw— Captain Mark Whitaker's body on the ground, burned and blistered. Whitaker's shoeless foot is unrecognizable from the burns, and his clothing is singed in multiple places, exposing his charred skin. His eyes are open wide, showing a look of fear, as if his final moments were terrifying. As I continue looking at the rest of his body, I see that his right hand is tightly gripping pliers that are missing their rubber grips.

"He must have been trying to fix these breakers when the second lightning strike hit the ship. That's the only thing that could have electrocuted him this badly," I say, doing my best to keep the sadness from my voice. "Let's head back up; none of these breakers will work now. The entire panel is wrecked."

When Tyler starts to walk away without a word, the glow of

my flashlight reveals a tear rolling down his face. Uncle Mark was the closest thing to family Tyler and I had left, and now he's gone too soon.

As we walk out of the engine room, I know my life will never be the same. If I know my brother, I know that what we just saw will be burned into his memory forever. Without a word, we head back to the atrium.

As we walk up the stairs, I notice the traffic has slowed significantly. There's no one screaming and yelling up and down the stairs. Now, many seem to know exactly where to go, walking with purpose. I tell the ones who still look a bit dazed and confused to follow Tyler and me to the atrium.

"None of these people have any injuries." I whisper to Tyler.

I wonder if these passengers were in their rooms during the chaos. Everyone we direct to the atrium is also wearing a life jacket, and most likely thinks the ship is sinking or that they are getting into a lifeboat.

As I enter the atrium, I spot Julie to the right, struggling to move a table. A split second later, a few more people walk over to help. Everyone is working together carefully to clean up the atrium. To my left, I see Logan holding himself up against a wall. Towels and chairs are scattered throughout the room, and each is occupied by an injured passenger waiting for help. It's been an hour or so since the explosion, and most people's emotions seem to have settled. However, some passengers are still sobbing uncontrollably, and a few are arguing. I also notice that most people have taken off their life vests and stacked them off to the side.

Tyler walks up beside me while I'm observing the room. I can tell he is shocked by his expression.

"I think these are all the passengers left on the ship," I say quietly, taking a deep breath. I do my best not to have a panic attack.

Tyler scans the room. He is quiet for a moment, watching everyone move about and helping each other. "You might be right. How many do you think are in here?"

"I'm not sure. From what I can see, I'd guess over a hundred." It's nowhere near the number of people who boarded the ship hours ago. The thought sends a chill down my spine, and I realize someone has to take charge, or else complete havoc will soon be unleashed again.

11

THEO

I grab the megaphone and turn on the siren for a few seconds to get everyone's attention. When everyone is staring at me, I feel a tinge of stage fright. I shove the nervousness down and tell myself these people need a leader, not a coward. I think back to all the times my dad would scrutinize me— *'Stand up straight, Theo. Stop slouching, Theo. Say yes, ma'am, and no, sir, Theo. Be a leader, Theo. Take charge, Theo.'* I shake my head and swallow the lump in my throat. *I am a leader, and I am the leader these people need right now.*

I bring the megaphone to my mouth, and when I open my mouth to speak, nothing comes out. All I can do is stand there, motionless, while everyone stares at me. I look around the room and see so many passengers injured, bruised, and bleeding. Many are slouched over, while others do their best to keep from crying— fear evident on everyone's faces. Then my eyes land on Julie and Tyler. I watch as my wife stares at me, her arms crossed over her chest. I look at my brother, and my mind races to the sad little boy who was always in the background when our father was being so hard on me and not giving a damn about him. He was always the one I could look to when we were younger, and he would always tell me what I needed to hear; he'd always have my back, just as he is doing right this second. As I stare at Julie and Tyler, I watch as Tyler mouths the words he has told me countless times— *You've got this, Theo.* And just like that, all my self-doubt disappears, and a jolt of confidence runs through me.

"Can I please have everyone's attention?" I say nervously, even though everyone is already staring at me. "Officer Logan, could you please make your way to me in the center of the atrium?"

In the far corner of the room, I watch Logan stand and hobble his way over to where everyone can see him. I take another deep breath and press my lips to the megaphone once more. "Hello, everyone. My name is Officer Parrish, but you can call me Theo. This is Officer Smith, but you can call him Logan."

Logan surveys the atrium, then looks back at me, and I continue. "We both manage operations on the ship. I want everyone to know I'm sorry for everything that's happening. We weren't prepared for the storm because it was an anomaly that the radar system didn't predict or detect. As you see, it caused significant damage inside and out, but fortunately, not enough to cause the ship to take on water or put it at risk of sinking. If you're still wearing a life vest, it's safe to take it off now, but if you're not comfortable doing that just yet, feel free to keep it on. I also went down to the engine room and found that the lightning strike to the ship caused a major power surge, destroying all the electrical wiring and infrastructure. The damage is so severe that we've lost all power and control of the ship. What this means is we are drifting freely. All the lights that are on now are running off generators that were only designed to last temporarily."

A wave of voices erupts and one man, louder than the rest, yells, "How are we going to get home? Where is the captain?"

A younger woman wearing sunglasses and holding a large purse stands up in the back. "When we get off the ship, my attorney will be suing you and everyone involved for this mess!"

There is a brief pause when she stops and before I can speak, more passengers begin shouting, asking similar questions and making threats. I feel my heart race and heat build in my face as all these people scream at me. I take a breath and block out the shouting. I quickly regain my composure, raise the megaphone above me, and start the siren again. Everyone stops shouting and stares at me.

"Everyone, please calm down. My late father, Benjamin Parrish, was the founder and owner of Parrish Cruise Line. He hand-selected our officer team, which includes Officer Logan and me. After his recent passing, I am now both an officer on this ship and the owner of this cruise line. I know everyone is under tremendous stress. Many passengers have lost their lives, and many more have fled during the chaos. To those of you who have lost loved ones tonight, I am truly sorry. I am no stranger to loss and have also lost a loved one tonight. Captain Whitaker was in the engine room trying to assess the damage and restart the engines when the second lightning strike hit the ship, killing him and completely frying all electrical systems on the ship." The entire room gasps. "If we all want to survive, we will need to work together. So, please, if everyone can keep quiet for a moment, we can work on a plan to help the injured and get rescued. Regardless of personal opinions, this is our reality, and we need to figure out how to survive."

The crowd begins to settle, and I see heads nodding in agreement. "Right now, Logan and I are the highest-ranking survivors on the ship—everyone, take a moment to look around. All of us may be the only people still left alive. Logan and I will lead the effort to develop a plan for our survival. To start, please raise your hand if you have any medical training. Are there any doctors here with us?"

Logan nudges me and leans in close to my ear. "We have a group of passengers with medical experience helping the injured over in the corner."

I pick up the megaphone again. "It's just been brought to my attention that we have a group of medical personnel already in place. If you are able to help tend to the injured, please do so in any way possible. That said, the ship's medical center is on the 2nd floor near the crew cabins. This floor isn't completely finished and is partially blocked off. I would appreciate it if a few of you would volunteer to carefully go down and gather all the medical supplies and equipment. Because the generators are only designed to last for a week, we will need help transferring the injured passengers and all the medical supplies to an area with windows for lighting and doors that open to the outside for fresh air. Is there anyone who can help with that?"

A moment later, hands start rising and people begin offering to help the injured and volunteering to gather supplies. "This ship is large, so we should all stay together and agree on the areas we want to use. If anyone has ideas about where things should go, now is the time to speak. Since the medical area needs fresh air, I suggest moving the medical center and the injured passengers to Hideout Cove—the spa and gym area on deck 11. This area has plenty of windows and several doors that open to the outdoors. Before we do that, we should check to make sure the area is safe. The initial explosion happened on decks 12 and 13."

I can't help but smile when I see people listening to me and the number of people who offer to help. "Okay, this is a great start. I'd also like to gather a group to go to the different floors and collect all the available food and water. Are there any crew members here who work in the kitchen?" I ask, then wait patiently for anyone to answer. It takes a moment before anyone responds. Then, when I think no one is going to volunteer, I see a hand go up in the back of the room. One man, who stands taller than the rest, has his arm raised high. I notice his arm is covered in black tattoos in different styles. He is also covered in splatters of blood. He is broad in the shoulders and reminds me of the actor Gerard Butler in the movie 300. "Hello," I say. "What's your name?"

I'm not surprised that the man's voice matches his physique – powerful and commanding. "I'm John. I work in the kitchen. What can I do to help?"

"Hey, John," I respond, standing up a bit straighter. "Thanks for volunteering. We can use all the help we can get. Could you gather a group and bring all the food and water from the kitchen?"

"Absolutely, I saw a few of us from the kitchen in here. I'll gather as many of us as I can. I suggest we turn one of the dining rooms into a food area, using the existing tables, and store all the food in one of the kitchens. Even though we don't have a way to cook, we could at least all eat in the same space and have plates and utensils." He says it in a tone that makes me believe he has been in charge before.

"Great idea. Let me know if Logan or I can help at all. Which dining area did you have in mind?" I ask.

John stands there silently for a moment before answering. "I'm not sure which dining areas aren't damaged. If the Oceanside Buffet is intact, I'd recommend it. It has the largest layout on the ship, with both indoor and outdoor seating. We can open the doors for fresh air, and the buffet-style stands would be great for serving food to everyone."

I lean over to Logan. "Thoughts?"

My eyebrow shoots up when Logan answers. "Make John a leader and put him in charge of inventory and gathering food."

I turn back to face John. "Thank you for stepping up and taking charge. We can't handle everything on our own. Are you willing to take on overseeing the food and water? We can all work together to figure out what we have and how to portion it out to make it last as long as possible."

John pauses for a moment, glancing at his fellow crew members, who smile and nod in approval. "Can I choose my own team to help me?"

I look back at Logan, and he nods his head in approval.

"Sure."

"Then yes, I'll do it," he answers.

After that, it occurs to me that we should also put someone in charge of the medical volunteers. I look at Logan, who is looking out at the crowd, stone-cold and unmoving. "Should we also appoint someone as the head of medical?" I run a hand through my hair. "This is far too much for me to handle. Unless you want to be in charge of it?" Logan is quiet for a moment as he looks around. I'm almost positive he's going to tell me he'll be in charge, since he likes to press the issue that he was the highest-ranking officer when the captain was alive.

"No," he says, to my surprise. "Someone else should do it." Logan must see my surprise because he instantly snaps back at me.

"What? Why are you looking at me like that?"

I shrug. "I figured you'd want to be in charge."

"No, I don't have that much training in the medical field. When I was in the military, I quickly learned the whole medical thing isn't for me," he responds, absently rubbing his leg.

I think back to my earlier interaction with the nurse. I kick myself for not remembering her name. I look over toward the group of medical people, and when I spot her standing with someone, the memory comes back. Mary. Her name is Mary.

I lean into Logan again, doing my best to keep my voice low. "There's a retired nurse I met earlier who might be willing to help. You think I should ask her?"

Logan is silent for a moment before he responds. "I'd check to see if there were any doctors who'd like to do it; if not, then yes, we should ask the nurse."

I pick up the megaphone again. "Mary, the registered nurse I met earlier, if you have a moment, could you come over? Also, if there are any doctors in the room, could you please join us as well?" *If my dad was on board, there would've been plenty of doctors.*

I watch Mary stand up and say something to the younger man beside her, pointing to a passenger lying on the ground. I assume she's telling the other guy what to do while she's gone. Then she walks toward Logan and me, and I can't help but notice she looks exhausted and is covered in blood. Her clothing is now stained even more than earlier when she introduced herself to me. I look down to see that her hands are also caked with fresh and dried blood.

Before she can speak, I lightly place my hand on her shoulder. "I'm sorry if I came on too strong earlier. I wasn't trying to be rude or disrespectful when I handed you the bag of supplies. Thank you for your help."

Mary wipes her face with the back of her arm, careful not to use her bloody hands.

"I understand," she says in a breathy, exhausted tone. "You were stressed, and this is an impossible situation we're all in."

I look around and notice that no one else has approached us. Mary is the closest thing to a doctor on the ship.

I look back at Mary and see a glint in her eyes that tells me she genuinely cares for people and wants to help them.

"Everyone here really appreciates how you've stepped up and taken charge. We'd like for you to manage taking care of the injured here on board. As far as we know, you're the most experienced here. You'd be in charge and could pick anyone you'd like to help you." I pause for a moment to let her process what I am asking. She doesn't respond right away. She turns back to look at the crowd. I wait a moment, then ask the question once more. "So, what do you say? Will you be in charge of medical?" She stays quiet a little longer. I watch her take in the entire room and from the look on her face, I can tell she is lost in thought. She isn't only looking at all the people. I watch as she looks around at the destroyed atrium. I see her eyes fill with tears, and she whispers something under her breath. I can't make out the words, and I don't bother asking. She slowly turns back to face me. Logan stands quietly next to me, patiently waiting for her to answer.

"I'd be honored to. Thanks for asking me," she answers.

I smile briefly, and for a moment, I'm tempted to hug her, but the blood covering her stops me. "Let's get through this together."

I look around again and realize everyone is still staring at Logan and me. Mary makes her way back to the injured, and I lift the megaphone to my lips. "Alright, everyone. We have a team of medical personnel in place, and another group is gathering all the available food and water. As I mentioned earlier, the ship's generator is only temporary. By any chance, did anyone board with a personal satellite phone?" The room is silent as I look around. "Okay, it was worth a shot. Unfortunately, we have lost all communication with the shore. All the flare guns have been fired, and every lifeboat was either taken or damaged while being deployed. With that in mind, we need to conserve everything we have. We don't have cell service, so I suggest turning off your phones to save battery. You'll need the flashlights at night, for when we lose the generators. Also, since there isn't any air conditioning on the ship since the power is out, I recommend everyone find a room with a balcony."

After my announcement, people immediately began congregating and talking loudly in the atrium. Slowly, one by one, I hear them turn to panic as the realization sinks in that all the survivors remaining are gathered here. I open my mouth to say something, anything that could calm everyone, but I think better of it. I've said what I needed to, and there really isn't anything else I can do at the moment. As I look around, I spot a young woman in the back of the room crying louder than anyone else. I watch as people around her try to soothe her. I watch as she shoves them away and begins to scream, then start hyperventilating. Others try to calm her, but nothing is working. I begin to walk over to see if there is anything I could do, but before I reach her, another woman approaches her and pulls her in for a long hug.

"Sweetheart, breathe, you're not alone," she says, reminding me of how my mother used to hold me when I was a child.

I watch as a sympathetic look washes over the crying woman. She drops to her knees and continues crying. After a moment, someone shouts my name. I go back to the center of the room, trying to figure out who called out to me, when I hear it again. I press the siren on the megaphone once more. This seems to be the only thing that gets everyone's attention. I watch as a man waves his hand for me to look at him.

"What should we do about all these bodies? We can't simply leave them here to decay," he says. His words once again stir strong emotional reactions from the crowd. Suddenly, the smell of blood and body odor consumes me. *How did I not smell it earlier?*

I blare the siren again. "You're right, but there's not much we can do right now. I suggest we move the bodies to a designated spot and cover them with sheets. If you wish, we can set up a prayer room for those who would like to say goodbye or pray. What do you think?" I ask. The room falls silent for a moment, with only muffled cries and sniffles. I watch as the crowd begins to nod in agreement. "I know this won't be easy, and some of you might find it hard, but if you're willing to help move the bodies, please gather in the middle of the room," I say, heading that direction myself.

After a moment of silence, people began to move toward the center of the room. One by one, several men and women joined them, their heads bowed and their faces showing sorrow.

"Thank you for volunteering to help. This isn't going to be easy, and I want you to know we really appreciate it. I'll head to the 12th deck to grab the backboards from near the pool. Hopefully, that will make moving the bodies easier." No one says anything. They are all standing and staring at me. "Does anyone have any questions?" I ask.

A young man dressed in a tattered shirt and bloodstained shorts raises his hand nervously.

"Yes?" I ask. The crowd turns to see who I'm addressing. He blinks as all eyes land on him.

"Are we all going to die on this ship?"

I pause, thinking, my gaze sweeping over my wife and brother. Before I answer, I feel my shoulders drop.

The room is so silent that I could hear a pin drop. I lower the megaphone; the room's quiet enough for them all to hear me. I take a deep breath and address everyone. "I'm not going to sugarcoat this, and I'm sorry, but I don't have an answer because I truly don't know. This isn't what we expected. We're trained for different scenarios, but this is unlike anything we've prepared for. All our training and practice scenarios point to one outcome: either get help through communication or abandon ship in lifeboats. Neither is an option now. All I can do is hope the storm hasn't pushed us too far off course and that someone on shore realizes we've lost contact and sends help." I pause, trying to sound confident while internally freaking out.

No one is looking for us... We never got the final approval for this practice voyage. The only people who know we are out here are friends and family.

I continue on. "Honestly, I don't even know where we are right now. I've never seen such a powerful storm before. All I know is that we're all we've got right now, and we need to stick together to survive."

There's a change in the air. People exchange glances, and everyone can see the sadness, heartache, and defeat on each other's faces. Then, something incredible happens: the entire room falls silent, and hands begin to come together. A few moments pass, and suddenly, every person in the room is holding someone else's hand, united. I look around again, my eyes filling with tears and my hand gripping Logan's, his gripping the person next to him. I close my eyes and whisper a quiet prayer to the universe, "*Please let us make it through this.*"

After a brief pause, we all release each other's hands. A moment later, the hope vanishes, and the room's mood turns somber again.

I can tell that everyone in the room is exhausted. I know everyone needs some sleep, or they'll be no help when it's needed most. "I can tell that you all want to help, and it's clear that emotions are running high right now, but if you're not working with the medical team or helping move bodies, please go back to your room and try to get some sleep. I'll be up for a while if anyone needs me. Remember, the 2nd, 3rd, 7th, and 8th floors are incomplete. Those floors are off-limits except for getting medical supplies from the 2nd floor. I've seen them myself, and they're not safe. The floors were initially locked to prevent access, but after the power failure, the system unlocked all the doors. So, please stay away from them. Thank you. Try to get some rest." I was right in my assumption that everyone is exhausted. One by one, the room becomes less crowded as everyone begins to leave.

Logan approaches me from behind and taps me to get my attention. "I can stay up and help if you'd like to get some sleep," he says.

"No, I'm okay. I can't sleep right now, even if I wanted to; my mind is all over the place. But thank you. How about you get some rest?"

"Appreciate the offer, but I feel the same way. I don't think I'll sleep anytime soon. I'm not even sure my mind has caught up with what's happening," Logan answers, then limps away, busying himself with cleaning up.

Tyler walks up with Julie not far behind. "Yeah, I agree. I can't sleep either. There's too much going on, and too much is happening. I'm going to stay and try to help around here."

"Me too," adds Julie.

"Thanks, that would be helpful," I answer. Taking a deep breath, I rub the back of my neck. Julie comes closer and wraps her arms around me. I wrap my arms around her, drawing her closer to me.

"You've got this," she whispers as she gives me a tight squeeze. I take another breath, exhaling a bit of the tightness in my chest... because I can't shake the feeling that none of this was an accident.

12

THEO

One of the men who volunteered earlier to help move the bodies approaches me, Tyler, and Julie. He stands about 6 feet tall, with shaggy blond hair and light eyes. His white shirt hugs his biceps and chest. He politely cuts into our conversation. "Theo, right?" he asks directly.

"Yeah, what can I do for you?"

"It's about moving the bodies—where should we take them? And where are the backboards you mentioned earlier?"

"I'll go get them for you. Give me a few minutes. Thank you for doing this. I don't think I caught your name?"

"It's Blake," he answers. "And it's no problem. I'm on the ship because my sister, Morgan, is a performer in one of the shows. When I last spoke with her, she was getting ready to meet me for dinner. She said she'd be late and would catch up with me in the dining room. I'm hoping to find her...alive."

Tyler, Julie, and I all look at him with sympathy.

Julie reaches out and gently rubs his arm. "We'll find her," she says, trying to offer him a glimmer of hope.

A small smile spreads across Blake's face. "Thanks, I pray you're right," he replies. "I don't think she would've risked getting on one of the lifeboats without me; maybe she's hiding in one of the rooms or is trapped somewhere."

We all stare at him with blank looks. The thought crosses my mind that his sister is probably gone for good. "Hey, why don't you come with me up to the top deck to get the backboards and we can search for your sister together?" As soon as I ask him, his face lights up.

"That's a great idea, thank you so much," he says.

"No, thank you for helping us. I hope we find your sister safe and sound somewhere," I say, knowing the chances of finding her alive are next to nothing. At least not on the top deck. "Want to head out now?"

Blake replies without hesitation, "Yeah, let's go."

From the back of the room, a woman rushes forward, straight toward us. I'm not sure who she is or where she's going, so I nudge Tyler, wary as she approaches. Tyler turns, and a smile spreads across his face. He must know her.

"Kelly!" he shouts.

I have no idea who this woman is, but clearly, he does.

"Who is she?" Julie whispers in my ear.

"I have absolutely no idea," I whisper back. Tyler opens his arms wide, and she rushes in, and they hug for what seems like an eternity. Julie and I look at each other, shrugging.

"I thought I wouldn't see you again. I can't believe all of this is happening," responds this mysterious woman. Her voice is a bit high and squeaky, and her eyes are bloodshot. She looks like she's been crying.

"Same here. This is crazy. I'm so glad you're safe," Tyler says as another woman and a man step out of the crowd and stand beside Kelly.

"I want you to meet my best friend, Sara, and her boyfriend, Chris," she says to Tyler, introducing the people who had just walked up.

"Hey, I'm Tyler. I wish we were meeting under different circumstances but now is better than never," he says. "It was great to meet you both, but I need to help Theo. I'll catch up with you all later once everything is settled." Then he turns to me. I'm not sure whether he wants to get away from them or whether I should introduce myself. I don't say anything. Instead, I just wave like an idiot, kiss my wife goodbye, and walk away.

I lead the way to the stairs, with Tyler and Blake following close behind. Most of the passengers have already gone to their rooms to get some sleep, while the rest of the volunteers are moving the bodies. The atrium is a complete disaster, with debris scattered everywhere. Someone has covered the bodies with sheets, tablecloths, or whatever linens they could find so others wouldn't have to look at them. The air is growing stuffier by the minute because of the lack of airflow, and soon the smell will be even more overwhelming. When we reach the stairwell, we find it mostly deserted. In the distance, I hear doors closing as the remaining passengers settle into what might be their new homes for the foreseeable future. I try not to think too much about how long we might be on this ship.

As we climb the stairwell, each body is draped with a white sheet. The sheets stand out sharply against the navy-blue carpet, crimson liquid seeping through the white cotton in varying amounts. With each body we pass, Blake stops and lifts each sheet, likely searching for his sister, Morgan. His face scrunches up with each view of the body, then falls flat every time it's not his sister. I do my best to keep my emotions in check, though it's very difficult to watch Blake. I'm shocked by how many bodies are lying covered. We've climbed five stories so far and counted at least 35 bodies.

Before we reach the door to the 12th deck, Blake turns to

Tyler and me, his expression pained. "This is going to be a long night."

"I really hope it's just the night and that we get off this ship come sun-up," responds Tyler. Together, the three of us stand in silence, lost in our own thoughts.

A moment later, Blake breaks the silence. "I'm sure it's only going to be one night. There's no way a ship this size is lost at sea. Surely someone will find us." *Yeah. If someone knew we were out here.*

Before I can stop myself, I blurt out, "Don't hold your breath. I wouldn't be surprised if they don't find us tonight. This ship is a needle in a haystack compared to the size of the ocean. But hopefully we're not as far from land as I suspect."

As soon as the words come out, I regret them. I watch as Blake's brows furrow.

"Suspect? What do you mean by that?" Blake asks, sounding a little on edge.

I rub the back of my neck and squeeze my eyes shut briefly, wishing I hadn't slipped up. I think for a moment, I need to decide whether to tell him the truth or what he wants to hear.

"Hey, what did you mean by that?" Blake asks, his tone becoming more agitated. "Is there something you're not telling us?" *Yes.*

But before I can answer, Tyler steps in and answers for me.

"I'm sure Theo didn't mean to make you think we won't be found. You're right, Blake. This is a large ship, and we're bound to be found."

Before Blake can get any more worked up, I decide to give an honest answer. "I didn't say anything earlier because everyone's emotions were already high, and I didn't want to upset them even more. I did some calculations in my head—they might not be perfect, but if I'm right, we're completely off our planned course. I suspect the winds were well over seventy-five miles per hour, which is why the ship was tipping so much and causing so much damage. The winds lasted a while, pushing us out to sea, as the swells were much larger than normal. So, honestly, I'm not sure when we'll be found. Like I said, if we're really that far out, we're a needle in a haystack." My words nearly bring Blake to his knees as he listens.

I watch his mouth drop open, and his expression shifts to surprise. Blake's eyes fill with tears. "This is why I haven't said anything; all that information would only upset everyone and might cause a riot." I look over to Tyler, and he appears equally shocked. "A riot is the last thing we need right now. Not to mention I'm not entirely sure about any of this. I'm just making a guess based on my training."

Blake's chest rises and falls as he takes a deep breath. He is silent for a moment, likely processing the information. "You might be right about your guess, Theo. Of course, if anyone knows, it's you. And I agree, we shouldn't share that information. It would be devastating for them. We've already been through so much in less than twenty-four hours. Let's not dwell on that now." He takes another deep breath, crossing his arms, then exhales. "So, where are we putting the bodies?

The question catches me off guard. Amid everything, I had completely forgotten to figure out that important detail.

"Give me a second to think," I say, trying to consider all the possible areas. *The only logical place is outside. But that's not going to last long.* I need to survey the area again before I can decide on a location.

The three of us step out onto the 12th deck. Tyler and I have already seen the devastation from the explosion, but Blake is seeing it for the first time. The rain is still falling, but not nearly like the torrential downpour it was earlier. The night sky is dark, and the clouds have cleared enough to let the moon's glow shine through.

I watch Blake's eyes widen in shock as he takes in the disaster around him.

"This looks like a war zone or a movie set," he says. I look around the deck and see that the rain has mostly washed away the blood, leaving only thin crimson streams flowing down the drains. Several bodies still lie scattered, their limbs torn apart. I notice right away that the place looks less chaotic than it did just hours earlier when it was engulfed in flames.

Out of the corner of my eye, I see Blake suddenly stop, his gaze fixed on something in the distance. I follow his line of sight and see several bodies on the ground. I'm not sure which one has caught Blake's attention, but I'm certain the only thing that would make him act like this is seeing his sister's body. Before I can ask him which body is Morgan's, Blake takes off running. I watch, frozen by his reaction, as Blake sprints across the deck. When he reaches the body that caught his attention, he drops to his knees and leans over her.

From across the deck, I make out the woman's long blonde hair, washed-out pink from the bloody water on the wet deck. Blake gently pushes the woman's hair aside and lifts her head. My breath is stolen at the sight of her face. A thick shard of shattered glass is embedded deep between her eyes. My eyes well up with tears. I understand how heartbreaking this moment must be for Blake. The deck remains silent, except for the relentless rain pounding on it. I'm startled when Blake lets out a scream. I don't interrupt him and let him have this moment to himself. It's not until after Blake has finished screaming that I walk over. Tyler, who is standing beside me, is motionless. I motion for him to stay back while I slowly walk over and place my hand on Blake's shoulder.

"I'm really sorry for your loss," I whisper. Blake doesn't

move or acknowledge my presence. I stand there, watching his shoulders rise and fall as he breathes fast and hard. Before I can say anything else, Blake turns to face me with a look that tells me he is furious and filled with hatred. He quickly stands and throws a punch. I feel a sharp pain in my forehead, then blood gushes from my face, and I stumble backward, completely shocked. *What the fuck?* Blake immediately climbs on top of me, pinning me to the wet deck. I do my best to raise my arms to shield my face. Blake grabs my arm, locking it under his leg. I shout to stop him just as his fist slams down on my mouth.

"It's all your damn fault," he shouts repeatedly. Just when I've lost all my energy to fight, Tyler tackles Blake. Their bodies land on the deck and slide in the rain. This time, Blake is the one pinned to the ground. I watch in horror as Tyler's fists pummel Blake's face. Frozen in shock, I can't turn away.

Tyler doesn't stop. He raises his bloodied hand again, slamming it down with each punch. "Don't...you...ever...touch...my... brother...again," he growls, each word punctuated with a punch. Tyler keeps punching, and I notice that Blake has stopped fighting back. I rush over and pull Tyler off. All the while, Tyler keeps shouting at Blake. This is a side of my brother I've never seen.

I do my best to calm him. "Okay, okay, he gets it, Tyler. You can stop now."

When Tyler finally gives up on going after Blake, we both glance over to see Blake slowly getting up. His face is badly beaten, one eye nearly swollen shut, his lip busted, and his brow bleeding. Blake slowly makes his way back to where we're standing, and Tyler starts shouting again, but I quickly shut him down, realizing Blake only reacted that way because of his discovery of his sister.

Blake slowly approaches Tyler and me. "I'm really sorry. I know it's not your fault. I just lost control. Morgan was my only family and my little sister; I was supposed to protect her," he says, dropping to the ground and sobbing.

I kneel beside him and rest my hand on his back. "I know there's nothing I can say that will bring your sister back, but I want you to know I'm truly sorry. I'm sorry this happened at all. I promise you I'll do everything I can to get us home."

After a moment of silence, Blake lifts a hand to wipe away his tears. He looks over at me, then at Tyler, his face a mask of sorrow.

"I know you mean that, and like I said, I'm sorry for lashing out. It's not going to help anyone if we start fighting."

As he speaks, Tyler steps in and stands beside me. "I'm sorry too. Just like you'd protect your sister, I had to protect my brother. I agree that fighting won't help anyone. Let's work together to figure out how to get off this damn ship. And when we do, I swear I'm never setting foot on anything that floats on water again."

We all stand silently, staring at each other as the rain pours down. Then, out of nowhere, Blake bursts into uncontrollable laughter. I look over at Tyler, and we exchange a confused look.

Tyler leans in and whispers, "He must be out of his mind. Who laughs in a situation like this?"

"Maybe this is how he copes with loss."

Blake suddenly stops laughing and turns to face Tyler and me. "This must be some sick joke, right? Everyone can come out now. It's not funny anymore. I'm ready to go home," he shouts, looking around as if waiting for a production crew to walk in and tell him this is all a big prank. I turn back to Tyler, giving him a concerned look. Tyler must know exactly what I'm thinking. We both slowly walk toward Blake, trying our best not to make any sudden movements, knowing that anything could set him off.

I reach out and place a hand on Blake's shoulder. "Hey, let's go back inside and get some sleep."

The moment the words leave my mouth, Blake's entire mood shifts. He stiffens and stares at Tyler and me. He stands in the pouring rain, as if he doesn't understand what I just said. Tyler and I stand motionless, watching him. Blake slowly turns to see his dead sister lying on the ground just a few feet away. The realization that his sister is not coming back must be unfolding in his mind. He begins to walk toward Morgan's body and Tyler, and I stand there, watching Blake as if he were a wild animal, unsure of what his next move will be.

Blake turns to face Tyler and me, defeat and sadness evident on his face. "She was all I had. No one else in our family is alive. I promised her I'd always protect her, and I didn't."

Before I can understand what's happening, Blake reaches down, pulls the shard of shattered glass from Morgan's face, and in one quick motion jabs it into his own throat, then slides it across his neck. Blood sprays out, covering our faces and clothes. Blake's body instantly drops to the ground, landing right beside Morgan. Tyler and I are in complete shock, my mouth open in disbelief, Tyler staring at the blood covering his shirt, eyes wide, and his head shaking. We face each other, both of us speechless. Stunned.

A moment later, Tyler breaks the silence. "We need to get the fuck off this ship."

PART 2

OCTOBER 2012

FIVE MONTHS DRIFTING

13

THEO

The crisp October air fills my lungs as a gentle breeze blows through my hair. Leaning against the railing on the top deck, I stare off into the vast open ocean and take in the beautiful setting sun about to dip below the horizon. The scene is so stunning, it looks like a watercolor painting in shades of pink, purple, and orange.

Despite the gorgeous view, the top deck is a constant reminder of all the terrible things that have happened to us. Most of the remaining passengers refuse to set foot on this deck. So, on clear nights like this, I sometimes find myself wandering up here to watch the sunset and clear my mind. Behind me, I hear the muffled sounds of a woman crying. I turn my head slightly to see who it might be. I assume whoever it is prefers to be alone, which is why she decided to come up here. That's one of the main reasons I often find myself here. She has her head between her knees, and her cries drift away into the quiet space around us. Giving her privacy, I look away.

Although the top deck is where I go to find solitude, I try to avoid the aft section. That part of the deck is where my emotions often get the best of me. I call that section the ship's memorial. Shortly after the explosion, we had to decide what to do with the decaying bodies. We couldn't leave them out in the sun; the smell alone was overwhelming. Logan and I held a vote among the remaining passengers about what options we had. In the end, for health reasons, it was decided that there was nothing to do but dispose of the remains. Logan wanted to just toss the bodies and get it over with, but in the end, we all agreed to make the process as respectful and peaceful as possible. As bodies were dropped into the ocean one by one, passengers brought memorabilia of friends or loved ones and placed the items in a pile as they said their goodbyes. It was a meaningful gesture in its own way and helped many of us because most of us had things we needed to say.

Once all the bodies were cleared from the upper deck, the area felt eerily empty. That day truly united everyone. After the memorial, once everyone had a chance to say their goodbyes, we gathered the largest, heaviest debris we could find. Working together, we arranged everything we could find to spell 'HELP' large enough for a plane overhead to see. Logan thought it was a waste of time, but the passengers loved the idea. At that moment, I understood that making passengers feel needed helped their emotions. From then on, tasks were assigned to keep our space livable. Giving passengers responsibilities gave them purpose and reasons to keep going. Friendships blossomed, fostering a sense of community.

Running my fingers through my hair, I sigh and take a deep breath. The sun slowly sets, and darkness spreads over the ship like a blanket. The temperature drops a few degrees, reminding me of fall at home. I imagine standing in my yard, leaves rustling in the wind, the scent of damp soil, the crisp air hinting at winter's approach. My eyes shoot open when the woman crying behind me begins shouting into the open, begging for someone to come save us. Instantly, a knot forms in my stomach, reminding me of the secret I'm keeping. I'm part of the reason no one is searching for the ship. I should have said no to my father. But nothing good can come from telling anyone that my father paid off everyone to turn a blind eye and let the ship sail without proper authorization. "It's only a test run, and the route isn't far," he'd said. All I know is that if my secret comes out, all hope will be lost, and hope is the only thing keeping us alive.

Before it gets any later, I decide to head back inside. I walk across the top deck to the forward section of the ship and can't help but stop and stare at the completely obliterated bridge. While standing, I see Brent, one of the passengers, who constantly follows Logan around. The sky has become so dark that I can barely see him. When he moves, his face catches a glint of moonlight. I notice it's wet, like he's been crying. I want to say something to him, maybe to make him feel better, but as soon as he sees me walking towards him, he rushes back inside. Most likely to find Logan. Something inside me twists, giving me a bad feeling. Why exactly does he always avoid me? I can't help but think it has something to do with Logan. He and I are supposed to be partners working together to help everyone, but my gut tells me he's hiding things from me. Lately, we rarely agree on anything anymore, and it's like he's formed his own group. I push the thought aside, too exhausted to worry about what Logan is doing. I'm too tired to care about anyone right now.

What I really need is a strong drink to help me forget everything. Especially how Julie constantly wants to be left alone. I try to ask her if there is anything I can do, and she always brushes me off and tells me she is fine. She hasn't touched me in weeks and every time I initiate, she pulls away. I'm not sure what else I can do. I'm doing my best to be a good husband, a good brother, and the leader that everyone expects me to be. The leader my father would expect me to be. Even though I feel like my facade is cracking, I know everyone looks up to me and that is why I keep pushing myself. When I walk through the hallways, people always stop me to ask if I have figured out a way to be rescued. I never know exactly what to say, so I always say the same thing.

No.

After all these months on the ship with the same people, I should know everyone's name, but I often forget. I always have so much on my plate. Not knowing if anyone is searching for us is killing me, and sooner or later, I'm going to completely fall apart.

On the bright side, my brother seems really happy with Kelly, considering our situation. And Mary has completely changed the spa and created a new medical center. She has plenty of help and handles everything herself. I've offered to help in any way I can, but she refuses and reassures me that she has it all under control. Luckily, I haven't had any injuries that need medical attention and Julie's hand healed without any complications, so I just accept that she is fine and move on.

Making my way back inside, I head to my room. The balcony door is slightly open, and I'm met with the chilly breeze again. Julie is already cuddled in bed, bundled up tightly under the blanket. Her eyes are closed and she looks so peaceful. Quietly, I slip out of my worn-out clothes and crawl into our bed. Wrapping my arms around her, I pull her in close. Gently, I slide her hair off the back of her neck, give her a light kiss, and whisper, "I love you." She doesn't move. "I'm sorry this is happening to us." With that, I close my eyes and drift to sleep, hoping tomorrow will be better than today.

14

JULIE

It's the dead of night, and I'm sleepless once again. I step out of bed and onto the balcony. The cool breeze is refreshing. I do my best to be as quiet as possible, careful not to wake Theo. I take a seat outside—a nightly routine I've gotten into over the past few months. The clear night sky is lit by the bright moon, casting a reflection on the calm ocean. As the temperature drops, the humid air sends a chill down my spine.

I decide to take a walk instead of staying on the balcony. I quietly step back inside and head to the closet for my once-white robe, embroidered with 'PARRISH CRUISE LINE' across the chest pocket. I slide it over my clothes and reach into the nightstand for the small flashlight Theo gave me a few days after the explosion. I tiptoe to the door, then step into the dark hallway. Before turning on my flashlight, I remain still, listening for anyone around. The only sound I hear is Theo snoring, so I walk down the hallway to the stairwell at the end, wondering if anyone else is awake.

As I approach Tyler's room, I do my best to avoid eavesdropping and keep walking. Despite my best efforts, the sound of moaning reaches my ears, and I can't help but roll my eyes. Just as I predicted, Tyler found a woman and didn't waste any time getting her in bed. Tyler and Kelly hit it off right away after meeting on the first day—at least that's what he'd said. A couple of weeks later, they decided to share Tyler's room. Every night, I've walked the hallways and it's inevitable that I hear them make love which makes me think about Theo. He has changed so much. More and more he reminds me of his father, controlling and always ready to prove himself to someone.

Another moan drifts through the air bringing me back to reality. Unfortunately, I have to walk past their room to get to the stairwell. Before turning on my flashlight, I trip over a plastic bag filled with trash. The bag rips open, and an empty champagne bottle rolls out, hitting Tyler's door with a loud thud. The moaning in the room suddenly stops, and then I hear footsteps approaching. There's no escape other than to run, which would be suspicious. *Dammit.* Tyler opens the door, wearing nothing but a towel. I point my flashlight in his face, temporarily blinding him with the intense light. He holds up his hand blocking the light.

"Sorry, I didn't see the bag on the ground. I didn't have my flashlight on," I say.

"It's okay. What are you doing out here in the middle of the night anyway?"

I shine my flashlight down the hall, then back at Tyler. "I was just trying to stretch my legs. I can't sleep – I'm sorry, I didn't mean to... interrupt."

Before Tyler can respond, Kelly walks over, wrapped in a sheet. "Hey, girl," she says in a cheerful high-pitched voice, like we are the best of friends. "Don't apologize. We weren't sleeping anyway."

Tyler laughs and I feel my face heat up. *This is so fucking awkward.* If there were a mirror, I'm sure I'd see my face as red as a tomato.

"Okay, well, that's my cue to leave. I'll see you all tomorrow. Have a good night." I turn on my heel and make a getaway for the stairwell.

"Have a goodnight, Julie, see you tomorrow," Kelly says cheerfully as Tyler shuts the door.

I take the stairs down to the 6th floor, my footsteps silent on the soft carpet. When I reach the library, the glass doors are closed, as always. I remind myself I'll need to find something to prop the door open. As I expected, the smell of mildew hits me the instant I open the door. Glancing around, I notice spots beginning to form on the books from the humidity and lack of airflow. I search the room for something I can use as a doorstop. I find a hefty wooden chair off to the side and decide it will suffice. Hopefully, the opening will prevent the mildew from getting any worse. Just as I always do, I go to the shelf of books by my favorite author. A twinge of regret crosses my mind for not propping the door open sooner when I see one of my books has spots along the edge.

After rifling through the stack of books, I find the one I was searching for and open it to the page I had marked. The stack of folded papers I had hidden is still there. I bring the stack of books to the porthole window, the only place I really enjoy sitting. It's quiet, secluded, and most importantly, it's not cold like when I try to sit outside. So as long as the mildew doesn't get any worse, I can keep this as my little hideout. The window is a little high up, so I have to leap a little. Once I'm settled, I pull out my flashlight and then change my mind about using any light. I'll need to save the batteries for when I actually need them. The moonlight is bright enough tonight so I can see the papers. After a few adjustments, I'm all nestled in, just how I like.

Carefully, I pull out the papers and unfold them. On top of

the stack is a piece of paper with Theo's name on it. I push that one aside and pull the second one. This page is blank except for 'Getting rescued plan' written in bold letters across the top. I pull out my pen to write, but I can't. My mind is blank; I can't think of anything that could help to get off this ship. One by one, I list all the things I've already checked: the lifeboats can't be fixed, and the fuel can't be used without working engines. The emergency equipment is gone. I think even harder. *What am I missing?* When an idea comes to mind, I jump out of my seat nearly falling to the ground and rush around the library, searching for *how-to* books, hoping to find one that can teach me how to build a boat. *There has to be a way to get off this ship.*

After a few minutes, I sigh and stare out the porthole window. I can't find anything here that helps. I hop back into the window and throw my head back in frustration. Staring out at the night sky, I grab the book with the hidden papers again. This time, I put the rescue plan sheet back inside and grabbed the one with Theo's name.

I gaze down at it and smile, and memories of my happiness with Theo flood in, taking me back over the years. Lost in faraway memories, my hand drifts to my stomach. I think about what it would be like to have kids. At first, I don't realize my finger is lingering on the small scar hidden under my shirt. When I realize I'm rubbing the scar, my happy memories vanish, and I'm brought back to the memory that has haunted me for years—the accident.

The losses I've experienced race through my mind. I think about my father and try to recall any good memories of him, but all I can remember is the car accident that altered everything. The memory of my mother's death takes over, and I struggle to remember what she looks like or the little things she used to say to me. The accident took both my mother and father from me. Even though my father survived, he might as well haven't.

I wipe the tears that roll down my face, and my hands go back to the scar on my stomach. That accident also robbed me of my future as a mother. My sadness quickly shifts to anger as the nightmare of that night plays out like a movie in my mind. I clench the paper in my hand, and it crumples in my fist. Tears continue to stream down my cheeks, and I take a deep breath, hold it for a second, then exhale. I remind myself that it has been a long time and that I can't let that horrible memory continue to haunt me. I smooth out the paper and sigh, tucking it back inside the book and closing my eyes, trying my best to relax. After a few deep breaths, my thoughts calm, and I'm finally able to clear my mind. Sliding down into a more comfortable position, I stare out the window and let my body and mind drift into a much-needed blissful sleep in the porthole window.

15

THEO

Sunlight floods the room, waking me. I stretch and yawn, then reach over to snuggle with Julie, like I try to do every morning. I'm not surprised she's not in bed. "Julie, are you in here?" I call out, thinking she's out on the balcony. There's no response. "Julie?" Still no answer. *Where is she?*

I grunt, not wanting to get up. Lying in bed, I feel the sun's warmth on my skin. The balcony door is cracked open a few inches and the cool breeze sneaking in every few seconds sends a chill through me. The wind whistling softly against the sliding glass door relaxes me. For a brief moment, my body is completely at ease, forgetting I'm stuck in the middle of the ocean. My calm is shattered when someone pounds on the door.

"Are you dressed?" Tyler asks. Before I have time to answer, I hear the door open and slam shut. "I'm coming in." His voice echoes through the room. I don't bother getting up. If he wants to talk, he can come find me.

Tyler makes his way over and sits on the bed beside me.

"Where's Julie?" he asks.

"I'm not sure. She was gone when I woke up," I answer, yawning and stretching. "Have you seen her this morning?"

"The last time I saw her was the middle of the night and she was in the hallway."

"In the hallway?" I ask, making sure I heard him correctly.

"Yeah. Kelly and I were up late last night, and we heard a knock on the door. When I opened it, Julie was standing there. It was weird, she said she was just taking a walk to stretch her legs, but Julie's weird so it really isn't that surprising."

"Well, it's not like she could have gone too far. Maybe she couldn't sleep."

"If I spot her, I'll let her know you're looking for her. Kelly and I are heading up to grab breakfast. Want to join?"

"No, I'll wait a bit and see if Julie shows up."

"Alright, suit yourself. We'll catch you later," Tyler tells me, shutting the door.

I roll back in bed again and gaze up at the ceiling. I think back over the past few weeks. *Julie hasn't been getting much rest lately.* I also remember her mentioning that she usually ends up at the library if she's not in bed when I wake up in the morning.

Reading before bed has been something she has done for as long as I can remember. Now that I think about it, she told me she prefers reading in the library window. She said that at night, when the sky is clear, the moonlight illuminates her book. She also said the balcony is cold and the window in the library doesn't open, so there's no breeze. The only way to find out is to get out of bed and go look for myself.

When I step off the stairwell onto the 6th floor, the library's glass doors are propped open with a chair. The closer I get, the stronger the smell of mildew becomes. I really hope Julie isn't reading in here every night. If she is, Mary is going to have her hands full, because there's no way this mildew isn't making her sick. As I walk through the library, I spot discoloration along the edges of almost all the books, a clear indication of how long we've been stuck at sea. Looking over at the window, I find Julie sleeping. I watch her sleep for a few seconds. Her strawberry-blonde hair hangs off the side of the window seat. She is sleeping so peacefully. Her chest slowly rises, then falls. *I wonder what she is dreaming about?*

On the floor beside her, there are books scattered everywhere. When I get closer, I realize they're books about famous disasters and biographies of survivors. Bending down to tidy up, I spot a single sheet of paper and I recognize Julie's handwriting right away. Picking it up, I read the bold letters 'Getting rescued plan,' with a list of checked items underneath. The list includes the things Julie and I have checked and rechecked multiple times: the lifeboats, communications, emergency gear, and so on. Then I see a little star scribbled with a small note: *"Find a book that teaches how to build a boat."*

I slowly stand up, holding the paper in my hand. I gaze at Julie sleeping. Seeing her cuddled up in the porthole window with the sunrise shining through makes me smile. She's so beautiful. As I walk toward her to wake her, I trip over a book, and she startles awake.

I watch her eyes look to the paper in my hand. Panic washes over her face. She springs up and then jumps down from the porthole window. As soon as she stands, she reaches for the paper and snatches it from me. Her eyes dart around like she is searching for something. I watch her eyes grow wide. I turn to look at what caught her attention, but before I do, she grabs me and pulls me in for a hug.

"I'm sorry. I haven't been myself, I've been driving myself mad trying to figure out a way for us to get off this ship," she says, holding me tight. I gently rub her arms to soothe her. She soon lets go, grabs the books stacked in the window, and turns to walk off. I step up into the window where she was sleeping and the second I turn around, Julie trips over the same thing I did, and her stack of books slams onto the floor.

"Are you alright?" I ask, leaping down from the porthole window to help. She waves her hands, shooing me away. A sheet of paper lies on the floor. When I reach out to pick it up, she snatches it away.

"I'm fine," she growls, and I take a step back. *What the fuck?*

She frantically picks up all the books like she's had shots of espresso and they're just now kicking in. *Did I do something to piss her off? Why is she acting like this?*

"It's okay to write what you wrote," I say, gently touching her shoulder.

She turns to look at me with a crazed look in her eye. "What did I write?" she quickly asks. I point to the paper in her hands. She glances down, reading it.

"The rescue plan you snatched from my hand."

"Oh, right," she says, turning back to pick up the remaining books.

"You don't have to hide it from me, babe," I say. "It's not a crazy notion or anything. We all desperately want to get off this ship."

"I know, but I don't want you worrying about me. I'm just trying to figure things out," she answers, a bit more calmly now that she has all the books pushed to the side and neatly stacked.

"I'm always going to worry about you. I love you. It's my job to keep you safe."

She doesn't respond right away. She looks away, her gaze landing back on the stack of books, then back at me. I can't help but notice the sudden sadness in her eyes.

"I love you too," she says softly as she pulls me in, nestling her head into the crook of my neck. We stand holding each other for a few seconds until she lifts her head and pulls my chin so we are facing each other. Slowly, she leans in and gives me a kiss. Her lips are soft when they touch mine. I feel my body relax at her touch. The mood is broken by a loud growl from my stomach, and neither of us can help but laugh hysterically.

"How about we go up and grab breakfast? I'm starving."

"Yeah. That sounds good," she says, setting the books down on a nearby shelf. I take her hand in mine, and we walk out.

After exiting the library, we climb the stairs, waving to people passing by whom we've gotten to know over the past few months. Everyone on the ship has now been assigned a job—making the ship a community. Every morning at sunrise, the hustle and bustle of the day begins. When Julie and I enter the buffet line for breakfast, we are instantly recognized. People start saying hello and waving as we walk by.

"Good morning, want to cut the line?" a voice calls out, but I'm not sure who said it, so I just respond in the general direction.

"No, no, we're fine waiting in the back of the line just like everyone else does."

"Good morning," John says as he approaches. His voice instantly grates on my nerves, and I can't help but regret putting him in charge months ago. At the time, we had no idea John was a power-hungry lunatic. Since taking over, he's been slowly undermining everything Logan and I have said and done.

Although, I have my suspicions that Logan and John are closer than they let on. On more than one occasion I've seen Logan and John whispering when no one else is around, and every time I've approached them, they act like they are discussing food or supplies in the kitchen. I never made a big deal out of it and let it go until a few weeks ago, when Logan and I had to meet with John after people complained that he was withholding food rations because someone said something that he didn't like. Logan acted like I was over reacting, and simply told John to not let it happen again. Over the past couple of months, all of us have learned that John wears his emotions on his sleeve and has a short temper. Everyone tiptoes around him, walking on eggshells, trying not to irritate him. It's frustrating.

I let out a breath, and Julie squeezes my hand. Julie knows just how much John has gotten under my skin, as I've kept her in the loop on the meetings Logan and I have had.

"Good morning, John," I say, a bit too cheerfully. Julie just stands next to me, and I catch a small smirk spreading across her face at my theatrics. "Looks like things are running smoothly here today," I add, glancing around at the full tables.

"Yeah, everything is going great," he says. "Thanks to me."

It takes everything in me not to say anything. Julie changes the subject when she sees my frustration.

"What's on the menu today? Eggs and bacon, French toast?" she asks sarcastically with a grin, trying to make a joke.

"You wish, sweetheart! You have no idea what I'd do for eggs and bacon. My mouth is watering just thinking about it," John answers.

"Yeah, same," I add, licking my lips without meaning to.

"Well, it looks like today we're having the same thing we've had every day before: oatmeal. Grab a packet of oatmeal and hot water from outside the fire pit, same as always," John says, walking away. Julie and I follow him. A few passengers are finished eating, and as we walk by, Julie and I pick up the empty dishes. John doesn't bother helping, like it's beneath him as he walks to the serving station. *Fucking conceited asshole.*

"Speaking of oatmeal packets, how's the inventory?" I ask quietly and set the dishes down in an empty dish tub.

"We're almost out. Honestly, I'm not sure what we'll give

these people when it runs out. We're going to have to start cutting meal portions soon," John says, scanning the room.

"Okay, we'll figure something out. But let's keep this under wraps for now. Who else knows about the situation?"

"No one other than Sara and Chris, and that's only because they're in charge of inventory," John says. *Good, the last thing I need right now is Logan to be up my ass, complaining about another issue I'll need to fix.*

"But Logan's going to figure it out. Look around. These people are already eating only two meals a day. Pretty soon, people are going to start getting sick from not having enough food," he adds.

"I'll talk to Logan. In the meantime, keep doing what you're doing. As for the people's health, I'll talk to Mary and ask what she thinks we should do."

At the edge of the room, a scuffle erupts as two men grab a young woman, pushing her to the ground and demanding she empty her pockets. *Is that Brent?*

"What's going on?" Julie asks as the commotion and yelling grow louder with each passing second.

"I'm not sure, but I'm going to find out," I say, running over.

"Get off her now!" I demand, grabbing the shirt of one of the two men pinning her. It *is* Brent. The woman slowly gets up and stands silently, her hands hidden in her pockets.

"What's going on here?" I ask the men sternly. Brent stares at me, unmoving. I can't quite read his expression; it's almost like he is thinking something directed to me, but won't say it.

Before Brent or the other guy can respond, John steps in. "I've been watching what's happening around here, and I've seen people taking their food rations with them. I told Jake and Brent to keep an eye out and stop anyone trying to take the food, but I didn't mean for them to tackle anyone to the ground." John gives the two men a fierce look. I watch them as John speaks and notice a look of fear cross Jake's face. Brent puts his head down like a scolded puppy. *Why are they so scared of him?*

"This is unacceptable. If the woman wants to take her food back to her room, why can't she?" Julie asks, with a bit of heat in her voice. One thing I've learned about my wife over the years is that when someone pisses her off, they'd better watch out. *Let's see what bullshit John comes up with this time.*

"We don't have enough food for people to hoard. They shouldn't be taking it out of the dining room. They need to eat it here or not eat at all," John answers.

"And who gave you the authority to make a decision like that?" I ask.

"You did," John responds matter-of-factly. "You and Logan gave me the authority when you both decided to put me in charge of the food and inventory." The look on his face is like he's daring me to challenge him. Instead of reacting, I simply remain calm and quiet.

"Well, I think—" Julie starts to say, but I stop her.

"Hold on, Julie," I say abruptly, staring at John with a silent warning. I take a step closer to John, our faces mere inches apart. The dining room falls silent as everyone watches us. "You're right. I did give you the authority. But we've already talked about this, John. You can't withhold food from anyone unless there's an issue. Logan and I have both told you to report it if you feel a change is needed, not to take matters into your own hands."

The room remains utterly still except for the howling wind outside the doors. John's face turns beet red, his eyes locked on mine. John takes a small step closer. He is a couple of inches taller than me, and he's so close to me that I can feel his hot breath on my skin. I don't move a muscle and hold my ground, refusing to be intimidated, waiting for John to say whatever it is he is going to say. The silence is suffocating.

John breaks the silence. "And exactly what are you going to do? Everyone knows you're all talk, and Logan's the one who's actually in charge. So why don't you take your pretty little wife and go back to your room before someone gets hurt?"

It takes every bit of self-control to keep my rage in check. I stand my ground, breathing heavily. I close my eyes, feeling the pressure in my ears build and the heat in my cheeks. I feel the vein in my forehead throbbing to the beat of my heart.

"You're completely out of line, John," I say as calmly as I can.

Julie gently grabs my arm. "Babe, let's go. He's not worth it."

I reflect on the past few months and on how John has been deliberately provoking me. I would propose one thing and John would go to Logan and argue that the exact opposite of what I said was needed. For instance, when I suggested assigning jobs that can be rotated each day, like dumping and fetching water for the makeshift bathrooms we created, to keep everyone busy and build a sense of community on the ship. John questioned my motives, implying I only wanted certain people to work, which wasn't true. When I stated everyone should have equal food rations, regardless of their gender, John disagreed, arguing that men needed double portions to maintain their strength, while women should have only one serving because they're smaller.

As I stand, staring back at John, it feels like time has stopped, the tension growing by the second. I have one hand in my pocket, gripping the cloth lining, fighting the urge to reach out and slam my fist into the ridiculous smirk on his face. I try to calm down by taking a breath in and exhaling, eventually agreeing with Julie. I turn and walk away without a word, knowing Julie is right; he isn't worth it.

"That's what I thought. Do what your wife says and walk away, you little bitch," says John.

I stop instantly, feeling every muscle in my body tense and Julie grips my arm tighter, tugging me.

"Theo...Please keep going. Just ignore him," she whispers.

Without thinking, I spin around and swing out my fist, making contact with John's nose. There's an audible crunch as his nose breaks, and he falls backward to the floor. Blood gushes down his face as I jump on top of John and repeatedly slam my fist down. I don't stop throwing punches until John's face is so badly beaten that he can't open his eyes. Tyler rushes up behind me, trying to pull me off, but I keep hammering down on John's face, elbowing Tyler by accident in the process. My body is growing weak from the energy I'm using on John. Suddenly, I'm knocked to the ground, and when I look up, I realize it's Tyler who tackled me.

"What the hell, Tyler?" I say, almost completely out of breath.

"That's enough. You're going to kill him," he growls. The room is still silent. Everyone remains frozen, all eyes on me.

I push Tyler off and stand up. But the shock of the moment gives way to horror as the group turns to John, who lies motionless on the floor. Seconds later, John gasps, taking in a deep breath, then begins coughing uncontrollably, spitting out blood. When he rolls over to face me, he spits out a tooth, and it lands directly at my feet. I kick the tooth back over to him, and then our eyes meet.

"I'm going to kill you, asshole. Mark my words. You and your whole damn family," John says, blood and saliva spewing with each word. Behind me, I hear movement and look to see Kelly step into view. Her brows are furrowed, and she has a look of anger I've never seen on her before.

She walks calmly to John, who is still sprawled on the floor.

I step back, letting her by. "What is this about, Kelly?" I ask, but she completely ignores me.

"Get up, asshole!" she screams. I'm stunned by her reaction.

Tyler begins to walk toward her, but I stop him by grabbing his arm. "Let her be."

John remains motionless on the floor, breathing hard. Kelly stares daggers at him, clearly frustrated that he's ignoring her. She pokes a finger in his face, leaning into his space now. "I'm tired of how you've been treating everyone, and it's going to stop right now. I said, get your ass up!" Drops of spit land on his forehead as she screams at him.

John looks up at Kelly and wipes his face with the back of his hand. "Fuck you," he mumbles almost inaudibly. "You can't tell me shit." Kelly's face turns a shade of red, and she kicks him square in the face, and another tooth flies out of his mouth.

What the actual fuck is happening? Everyone in the room gasps. I've got to stop this. "Wait, hold on, Kelly, that is enough..." I say. She doesn't pay me any mind. She leans over John again and grabs him by the ear. I stop moving to see what she is going to do next.

"Get the fuck up," she says, her voice dripping with venom.

"Babe!" Tyler says. "What are you doing?"

He sounds just as confused as the rest of us. Kelly looks up at Tyler and stares at him for a beat, frowning, then shakes her head and refocuses on John. She twists his ear viciously, and he yelps and jerks his head, scrabbling against the floor to get away from her. I have to put a stop to this.

"Ok, Kelly, that's enough, he's learned his lesson. You can let him go." I keep my voice calm and low, almost like I am trying not to spook a scared animal. Granted, I don't know the woman very well, but I've never seen her act like this before.

From the look on my brother's face, this is new to him too. It's almost like she is someone else entirely.

She looks at me, and I blink at the expression on her face. It's almost like she's staring through me. "This guy's not going to keep acting like he's some hotshot leader." Her voice is calm and deadly at the same time. "All this asshole does is talk down to everyone like he is something almighty. He needs to be put in his place." She's breathing slowly and steady, but the look in her eyes is almost feral, dangerous. No one in the room moves or makes a sound.

"I understand that," I say, stepping toward her. "But Logan and I will take care of that, okay?" From the corner of my eye, I see Tyler nodding in agreement, then I catch movement behind Kelly and see John rise up quickly, taking a swing toward Kelly.

"Babe, look out, behind you!" Tyler shouts. Kelly turns around and ducks just a second before John's fist sails past her cheek, nearly hitting her. Gasps sound in the room as Kelly rears back and then stands up straight. She looks at John, from his face to his fist, like she can't believe he tried to hit her. *I can't believe he swung at her.*

"Ok, John, you need to step away now," I cut in, and nudge Tyler on the shoulder. "Help me out, bro–"

"Don't you fucking threaten my family again," says Kelly in an unusually calm voice that sends a chill down my spine. *Family? Again?*

John looks at her then laughs, suddenly spitting blood across her face. "Get out of my face, you stupid whore." He shoves her to the side as he begins to walk back to the serving table.

What looks like a dark cloud passes over Kelly's face as she looks around her, scanning the table. Her eyes land on a fork and she grabs it, then runs after John. Everything seems to be happening in slow motion at this moment. Before any of us can stop her, she launches herself on his back, wraps an arm around his face, and viciously starts stabbing him repeatedly in the neck. John screams, grabbing at her from behind, yanking a handful of her hair as he wrenches his body, trying to throw her off. A few others in the crowd scream, and the sound spurs Tyler and me into motion.

Tyler grabs Kelly and pulls her off John, all the while she shouts, "Fuck you, asshole. I hope you rot in hel–"

Kelly is silenced when John's body flails and his foot connects with her head, knocking her out cold. John slumps to the floor, and I grab a few linen napkins from the table and press them tightly to his neck. Kelly's body is limp in Tyler's arms.

"Somebody get Mary!" I press my hands to the wound on John's neck. He stares up at me and the linen on his neck blooms into a deep crimson stain. "I'm so sorry, John. I don't..." I can't think of anything to say as he looks up at me, blood gurgling from his mouth. He blinks at me, then tries to speak, but cannot get any words out. He shudders out a breath, then slumps back onto the floor. I look up and see the others in the room, utterly shocked by what just happened.

Then, when I look over at Kelly, she's conscious again. Tyler is stroking her hair and speaking softly in her ear. She stares at me with tears in her eyes, almost with a scared look. Then she looks away, snuggling back into Tyler's arms. *How am I going to fix this?*

I look down at John's lifeless body. I'm too stunned to comprehend exactly what happened. I lay his body down and walk over to Kelly and Tyler. I gently grab Kelly by the hand. "Come with me." Her entire mood has changed from a few minutes ago. Then I glance over at Julie, who looks equally shocked. "Come with us, Julie," I tell her. "Tyler, can you track down Logan and Mary and meet us in the Stardust Lounge?"

"What are you doing, Theo?" Tyler asks, putting an arm out to stop me.

"Babe, I'll be fine," Kelly says. Her gaze falls on John's body. "Not again," she whispers so quietly I almost didn't hear her. *Again?* I don't say anything else to her. Tyler doesn't argue any longer, and he heads out the door.

I look at the crowd that is still standing, staring at John's body. "Can a group of you please clean all this up. Toss John's body overboard like we did with the others."

"She can't get away with this!" shouts someone from the back, followed by many others screaming too.

"We will handle this. I know everyone is upset, but we need to just–"

"Just what?" Brent asks from the back of the room.

"Just clean this up, then everyone goes back to their rooms or finds something to do. Logan and I will sort everything else out."

"I'm sure Logan will know exactly what to do," Brent says, a smirk dancing on the edge of his lips.

16

THEO

A couple of months ago, the leaders decided that a private meeting area would be a good idea. The Stardust Lounge stood out because it has doors that can be closed and windows that let in natural light. Additionally, it's the only lounge on the ship with access to a small, exclusive outdoor patio, which keeps the air fresh and prevents the lounge from becoming musty due to poor airflow.

Julie, Kelly, and I are seated at a large round marble table in the center of the room, waiting for Logan, Tyler, and Mary to arrive. The silence is unbearable. Kelly's leg bounces nervously, causing the seat she is in to rattle, which is going to make me lose my mind. Her face is streaked with dried blood, tears, and snot. Every few minutes, she pauses, and then when I think it's all over, that fucking leg starts up again. I can tell she is nervous because, in addition to the nonstop bouncing, she also keeps picking at the skin around her thumbnail. Julie reaches out to touch her leg, which is either an attempt to help calm Kelly down, or it's annoying her as much as it is me.

The touch must have worked because Kelly's leg goes still,

and suddenly Julie is holding her tightly as Kelly starts to sob uncontrollably.

As Julie rubs her back, I hear her whisper, "Everything's going to be okay." I can't help but think for a moment about how wonderful a mother Julie would be. She is so nurturing. A moment later, Kelly's face lifts. Her eyebrows shoot up, and her eyes dart to Julie.

"How can you say that? Look around, Julie—we're lying in our own grave, and nobody's going to dig us out. I just killed a man!" Before Julie or I can respond, the lounge doors fling open, and Mary, Logan, and Tyler walk in.

"What took you so long?" I ask, irritated. Tyler rushes over to Kelly, who is still in a state of panic.

"Hey, everything's going to be fine, I promise," Tyler says, gently wiping her face with a cloth. Kelly doesn't respond. She just sits there quietly; her blood-shot eyes fixed on the floor.

Mary speaks up, drawing my attention to her. "I was in the middle of taking inventory of our medical supplies."

"And I was dealing with a personal issue," Logan adds, glancing at Kelly. "What happened to her?"

"What kind of personal issue?" I ask, furrowing my eyebrows.

"If I wanted you to know, I wouldn't have said personal," Logan replies in his usual assholish manner I've become accustomed to lately. "Now, what happened to her, and what's this emergency meeting about? Tyler wouldn't tell us anything." I run my fingers through my hair and let out a ragged breath.

"We have a... situation," I answer, trying to find the right words to explain Kelly's actions.

"Clearly, or I wouldn't be here, so what is it?" Logan demands. Everyone in the room stays silent for a moment. "Well? I'm waiting. Are you going to tell me, or are we just going to sit here in fucking suspense all goddamned day?"

"Julie and I went up to grab some breakfast and —"

"Kelly became a badass and killed John with a dinner fork," Julie says abruptly, like it's something Kelly should be proud of.

Mary gasps at Julie's words. "The asshole deserved it," Julie replies.

"What the fuck? How did this happen?" Logan asks.

"It's a long story," I say.

"Make it short and tell me. We can't have people killing people without consequences; the whole ship will be in an uproar." Ignoring his snide comment, I look to the back of the room and see Mary standing with her hand still over her mouth. She glances around the room, then back to Kelly.

"Are you okay?" she asks.

Kelly doesn't answer. Tyler speaks up instead, "Yeah, she's fine."

"Once again, can someone tell me what the fuck actually happened?" Logan demands. His voice is growing more irritated by the second.

"Julie and I went to breakfast like we always do," I say as I rub my hand over my beard. "John walks up, trying to make small talk with us. Suddenly, there are screams, and Jake and Brent are pinning a woman to the ground, demanding she empty her pockets. I asked what was going on, and John said people have started taking their rations to their rooms to eat, and that if they didn't eat in the dining room, they won't eat at all. He then lost his shit on me and Julie after I told him he couldn't do that, then he basically threatened us. Then, from the back of the room, Kelly comes up and gets in his face; he's a dick to her as well, and after he tries to punch her, she loses it. She picks up the fork, jumps on his back, and stabs him in the neck multiple times. And now we're here in a meeting about what our next steps should be."

"Well, fuck me," Logan says, massaging his temples. Then

his face lights up with a smile, and he bursts into laughter. "I guess I should say thanks, Kelly. That guy was a total power-hungry asshole, and now there's one less thing to worry about. Honestly, I was ready to take him out myself."

Not the reaction I was expecting. We all give him a look of surprise.

"Logan, this isn't a joke. This is serious. She killed someone, and people are going to ask questions and demand she be punished. We can't let people just go around killing each other." I look over at Kelly, and she is still staring at the floor. "So what are we going to do now?"

"We don't have a jail to put her in. So what are our options?" Logan asks.

"We can take away her food rations for a few days as punishment," Mary says.

"No, I'm not letting her go without food," Tyler says.

"Well, Tyler, what do you suggest? She killed someone. Taking away food sure seems like a pretty light punishment if you ask me." He doesn't answer, and I can tell he is contemplating what to say. "Or, we could barricade her in a room like a prison cell. Would you rather us do that?" I ask. Looking around the room, no one else says anything. They all just stare blankly.

"I'm fine without eating. I really am sorry." Kelly says.

"Sorry doesn't bring John back, or prevent the rest of the passengers from acting out," I snap back. Julie grabs my hand and I pull it away.

"I actually like the idea of building a prison cell," Logan says, grinning like all of this is some kind of sick comedy show for him. *I can't fucking stand this guy.*

"No. We are not locking her up. She will go without food for three days." I quickly change the subject before anyone else can object. "So, who's going to take over John's job? We need someone we can trust and who won't try to control everything."

"You're right. That can also be part of her punishment. Since Kelly is the one who killed him, she should take his responsibility," Logan answers, his gaze landing directly on Kelly. "Kelly, you're going to be the one to take over John's job, and then you can explain to each person you see why you did it."

"Okay," Kelly says quickly, as if trying to change the subject. I stare at her, studying her, reading her lips as she whispers under her breath, "I can't believe I did this again." *Again?*

"What did you say, babe?" Tyler asks.

"Oh, nothing," Kelly answers. Tyler looks at her, puzzled, but she just smiles at him. This is the second time she's alluded to this happening before—I'm going to have to get her alone to figure out what she meant.

Tyler looks at her for a moment longer, but doesn't push her. Instead, he sits back in his seat and lets the situation go. I can't stop thinking about what she said. *Again? Does this mean she has killed someone in the past? I need to find out more about this woman before my brother is in too deep. I'm the only one who cares enough to protect him.*

I shake the thoughts from my head and bring my attention back to Logan. "Did John mention anything to you about the inventory running low?"

Logan squints at me, and I swear for a split second, he looks puzzled that I asked. "He told me the other day that the kitchen was low on oatmeal, but that's it. Did he tell you anything else?" He answers quickly.

I stare at him for a moment as something occurs to me: John made it seem like he hadn't said anything to Logan about the food shortage yet. So how would he know that? Instead of calling him on it, though, I try a different tactic. "No, he didn't say anything other than that."

For a split second, again, Logan looks at me strangely. Only this time, it's like he doesn't like my answer, then his face goes blank. It makes me wonder if he was the one who told John to keep an eye out. Logan is sneaky as hell, so I really wouldn't put it past him.

"We have another issue on our hands," Mary says, and we all look in her direction. *Oh, for fuck's sake, we can't handle any other surprises right now.*

"What is it?" I ask.

"There are a few people who've started to get sick, and I have no idea why."

Everyone starts speaking over one another, all at once.

"What's going on?"

"Is it serious?"

"What kind of illness?"

"What are the symptoms?"

"Are we all going to get sick?" Tyler asks, and with that question, the entire room goes silent for a moment, then the questions start flying again. Mary puts her fingertips in her mouth and lets out a sharp whistle, and we immediately quiet again.

"I guess those years of softball have finally come in handy," she says, shaking her head. "Okay, give me a second, and I'll explain as best as I can." She takes in a deep breath and exhales dramatically. "It started a few days ago..."

17

THEO

Everyone stares at Mary, standing in the center of the room. We all wait patiently for her to explain.

"About a week ago, Jerry walked in to see me," she continues, nervously rubbing her arm back and forth.

"Jerry, the one that was helping you with patients after the explosion?" I glance over at Logan, then back to Mary. He'd been part of the original crew Mary put together.

She nods. "Anyway, he came in complaining of not feeling well. His symptoms were classic flu symptoms: fever, night sweats, cough, sore throat, and lethargy. At first, I didn't think much of it. I asked him if I could check his urine, and turns out, he was severely dehydrated. I gave him some water and told him to rest. The following day, I checked again. Nothing changed, so I set up an IV with a saline drip to rehydrate him. I only had a few of the saline bags, so I figured I'd give it a try." I don't realize until I hit the quick of my nail that I'm picking at them while listening to Mary, a habit I've had since childhood when I'm nervous or stressed.

Logan interrupts, "Okay, so he has the flu. No big deal. Just give him meds and tell him to rest. We all know how to handle that. Let's get back to—"

Mary cuts him off. "Look, I don't tell you how to run the ship, so don't try to tell me how to run the medical center. Unless you want to take over?" she says, pointing a finger at him. I smile slightly, loving seeing someone puts Logan in his place.

Logan stares at her for a moment. "Okay, point taken. Continue." Then he takes a step back.

I glance around the room, noticing I'm not the only one who's nervous: Kelly's leg is bouncing again as she sits at the table. Tyler's rubbing her hand soothingly, sitting beside her. Julie is staring at Mary like she is studying her every move. Logan, after being scolded, has his arms crossed, irritated, and I can see his masseters expanding at his jawline as he clenches his teeth. He turns and catches me staring at him.

"What the fuck are you looking at?" he snarls. I don't say anything and refocus my attention on Mary and watch as she fidgets with her fingers, the same way I tend to do.

"As I was saying, at first it was classic flu-like symptoms. I treated him for a few days, and he seemed to be improving. I thought we were on the right track but things suddenly took a turn for the worse. Four other people came in two days ago with the same symptoms. I did the same thing, minus the IV drip, since those are extremely limited. After I got the other four people settled and resting, Jerry started complaining again in the middle of the night.

"This time, the symptoms were much more severe. He started screaming in pain, grabbing his abdomen. He rolled off the bed onto the floor, still screaming, begging for help. I tried to ask what was causing him pain, but he couldn't answer. Then, all of the sudden, he stopped screaming, and he let go of his stomach. He looked at me with a blank stare, and I asked if he was okay, but he didn't respond. A moment later, he vomited.

There didn't seem to be any food in it, nothing but blood and bile. He then wiped his mouth, looked up at me, smiled, and crawled back into bed, telling me that he felt much better.

"It was the strangest thing I'd ever witnessed. His skin had turned ghostly pale, and the whites of his eyes were bloodshot. Not a split second later, his eyes rolled to the back of his head, and he started seizing. He started foaming at the mouth. I did what I'm trained to do and rushed to turn him on his side, but when I got to him, the seizure had subsided. His body lay motionless, and he wasn't breathing. I couldn't save him." Her head falls, and she faces the floor at the last words.

The room remains silent, and Julie's hand is pressed to her face, covering her mouth. Everyone else seems equally stunned, unable to find their words.

Logan is the first to break the silence. "What the fuck, Mary? Why are you just telling us this now?"

"I thought I had it under control, and everyone else is busy with their own jobs. What could any of you have done if I had told you?" she demands, hands in the air.

Logan doesn't answer. Mary's right. None of us could have done anything differently.

I speak up, my voice soft. "Mary, what about you? Could you have been exposed to whatever he has— had?" I correct myself. "What about the others?" Everyone in the room panics, asking questions and desperately trying to put as much distance between themselves and her. It isn't the reaction I was trying to provoke.

"No. Everyone, calm down!" Mary demands.

We all go silent. "I don't think this is an airborne illness. All the people I'm treating have the same story linking them to the same point of contact." She lets her words hang in the air for a beat.

"What's the point of contact?" I ask.

"They all told me that John had given them extra food in exchange for doing his dirty work."

"What dirty work?" asks Tyler, taking the words right out of my mouth. Instantly, I shift my attention to Logan, watching his reaction and suspecting he and John had been scheming behind my back. Logan refuses to make eye contact with me. Instead, his eyes dart around the room, almost like he is trying not to fidget. My gut is telling me he knows something the rest of us don't.

"At first, they wouldn't say anything, like it was a huge secret. It wasn't until Jerry died that they finally decided to speak up," she says.

"So then, we should be good, right? You have all the people John was feeding?" Julie asks.

"It's not that simple," Mary responds.

"Well, sweetie, make it simple for us," Logan murmurs.

"First off, Logan, don't call me sweetie. Show me some respect." Her words strike a nerve with Logan. His face hardens, but he doesn't respond. "Second, I can't explain it. I have no idea what's causing this. I'm still trying to figure out what John gave them and what would cause so much blood in the abdomen. Also, within the next few days, if the other patients' symptoms match Jerry's, it's likely they'll all be dead sooner rather than later."

I push my chair back and stand from the table. Everyone instantly turns to me.

"Where are you going?" Julie asks.

"It sounds like we all need to take a little field trip to the medical center and figure out what they know," I explain. "We can continue the Kelly conversation later." My eyes land on Kelly as soon as the words leave my mouth.

"It's going to be fine. Stop worrying," says Tyler, putting his arm around her.

"What information are you hoping to get from them?" Mary asks.

"Anything. I want to know why they were doing dirty work for John." Mary doesn't get a chance to respond before Logan interjects.

"I'm sure they are just spewing bullshit. I know how to put an end to their suffering. You said you don't know what their sickness is, Mary? They're going to die anyway, so I'll speed up the process and put them out of their misery. Then the problem is solved, no more spies and no more John causing endless headaches. It's as simple as that. Now get the fuck out of my way." He doesn't wait for anyone to protest. He shoves Mary out the way and storms out the door.

Once the doors close behind Logan, Tyler breaks the silence, "He was joking, right?"

Mary looks over at Tyler. "No, I don't think so."

"I don't think Logan is the joking type," I say. "Let's go figure out what is happening and stop him before he does something crazy."

18

THEO

Walking into the medical center feels like a breath of fresh air. I used to visit here the first week to check on Mary when everything was a bit more chaotic, but as things settled, she told me I didn't need to keep checking on her, she had everything under control. I listened to her and haven't been back since.

There's an aroma in the air that I can't quite put my finger on, but whatever she has in here smells fresh and clean, like disinfectant cleaner, reminding me of a hospital. The place is pristine: The polished faux-wood floors are sparkling clean, a stark contrast to when they were covered in blood. There are no shades on the windows any longer, so the sunlight is no longer blocked. All the chairs are lined up along the wall, a setup that reminds me of a waiting room. There's a help desk just inside the door, behind it, and tan stones are lined from the floor to the ceiling. My attention is immediately drawn as Tyler walks in the door, obnoxiously loud as per usual.

"Whoa, I like what you've done with the place," he says with a nod, looking around. "It's been a while since I've seen this place."

Mary turns around, smiling. "Thanks, Tyler. I did what I could to make people feel comfortable while they're here."

"He's right, it does look really nice," I say, looking out the window at the vast ocean that stretches for endless miles. I look away, not wanting to think about how far we have drifted.

"Enough with the small talk. Where are the patients you were telling us about?" Logan asks, walking around.

"They're over here." Mary leads us toward the hallway that's lined with doors that were originally for the massage rooms. "I put the patients in separate rooms to keep them contained in case what they had was contagious, but, like I said, I don't believe it is. But given what happened to the first one, I wanted to be cautious."

"Hey, who else knows about this?" Logan asks.

"Just a few people who help me around here and there," she answers. "Why do you ask?"

"Because we need to keep this quiet," Logan answers. "We don't want anyone else finding out about...whatever the hell this is. Everyone would lose their shit if word got out."

"Finally something we can agree on," Mary says, turning the knob to the first room.

"Should we put on a mask or something?" Kelly asks, her nose scrunched up like she just caught a whiff of something rancid.

"No, I don't think a mask is necessary. But, if you feel more comfortable staying outside the room, feel free to do so. I understand your caution."

We stand together at the door housing one of the sick patients. As soon as it opens, my face contorts into the same expression as Kelly's. There's an overwhelming smell of days-old vomit, urine, and sour sweat that's so strong that I'm almost certain when I walk out of this room, the stench will be clinging to my clothing like nicotine from a small town dive bar.

"Holy fuck, that's awful," Tyler murmurs. Kelly, standing next to him, elbows him in the side. Good thing because at this point, I'm ready to punch him in the throat if he doesn't shut up. Everyone gathers around the sleeping patient, holding our noses to block out the stench. In the center of the room, a young woman lies asleep, her skin damp, and her hair clings to her face. She is completely unfazed by the commotion around her.

"This is Beatrice," Mary says quietly.

"She isn't contagious, right?" Logan asks.

"She has already told us the answer to that question." My eyes are locked on Logan's. He opens his mouth to say something, but then closes it and turns around and walks over to Beatrice.

"Wake up," Logan shouts, his hand reaching out to her shoulder. Mary intercepts and swats his hand away before he can touch her.

"Show some goddamn respect, Logan," Mary snaps, her tone stern, making it clear how much she despises him. Mary walks over and gently tries to wake Beatrice.

"Beatrice, wake up, honey. Someone wants to talk to you." Beatrice stirs but doesn't fully wake. "Beatrice, honey, I need you to wake up for a minute," Mary says, still gently nudging her. Beatrice groans and rolls over. She opens her eyes and sees Mary, then gives her a small smile, unaware that the rest of us are standing in the room. She lies there for a moment, her eyes fluttering to wake up. "Hey sweetheart. How are you feeling?"

Before Beatrice answers, she stretches and lets out a yawn and settles into the bed. I watch as she looks past Mary and notices me and the others standing in the room. But it's not until her eyes land on Logan that she reacts. When she sees him, her entire body goes stiff and her eyes widen.

She sits up quickly and looks at Mary, panic clouding her face. "Thanks for taking care of me," she says, running her fingers through her matted hair, staring down at the floor. "I'm feeling much better. I'm ready to go back to my room now."

Beatrice tries her best to get out of bed, but Mary, looking just as confused as I feel by the sudden burst of energy, looks around at everyone, then back at Beatrice.

"I don't think that's a good idea," Mary says, her voice slow and calm. "If the symptoms are the same as your friend's, this is only temporary relief. I need to figure out how to stop whatever's making you sick before it gets worse."

Beatrice begins to shift in the bed. "No, really, I'm fine. I'm feeling much better. Can I please go back to my room?" Beatrice is clearly getting more uncomfortable every time her eyes flick over to where Logan is standing.

There is no way I'm the only one who notices the sudden change in her attitude whenever her eyes meet Logan's. It's as if she's afraid to speak with him in the room.

"Hey, I'm Theo, remember me?" I say softly, stepping in front of her and intentionally blocking her view of Logan. "If I remember correctly, you live in... Colorado or somewhere out west, and you have a dog you rescued?"

"Wow. How did you remember that?" asks Beatrice, her shoulders dropping and becoming a bit more relaxed.

"I love animals, and growing up, I was never allowed to have pets but I always said I'd rescue a puppy one day. That's why I remembered you." I quickly bring the conversation back to what I came here for, hoping I've gained a little of her trust. "Ok, so I'm just trying to figure out what happened here. It's really important that we figure out what's going on so no one else gets sick. Can you tell me what happened?"

Her eyes lock onto mine like she's studying me. Then I watch her gaze shift past me to Logan. Instantly, her body tenses again. I sense she's about to clam up, so I shift on my feet a little to block her from seeing Logan and pulling her attention back to me. "Focus on me, Beatrice. Let's pretend no one else is in the room and it's just us." Tears well up in her eyes. I reach down and grab her hand, trying to comfort her. She's trembling and her palms are clammy. I know something is wrong. Very wrong, and it has to do with Logan.

"I need everyone to leave the room," I say, never taking my eyes off Beatrice. "Except you, Mary. Please stay."

"Sure," Mary responds.

"No, I'm not leaving." Logan crosses his arms and leans against the wall. "If she's going to talk, she can do it with all of us in the room." Beatrice flinches at the sound of his voice.

I speak softly, trying not to spook Beatrice while still getting Logan out. "Logan, please, just five minutes to talk? She's upset, and you're only making her more scared."

"No, Theo, this is serious. Besides," Logan says, pointing at Beatrice. "She doesn't look that sick to me. I don't give a rat's ass if she's scared. She needs to tell us what the hell happened."

Beatrice struggles to sit up in bed. I reach out to help her, but she lifts her hand to stop me. Then, once she is sitting up, she steadies herself and stands, staring directly at Logan. Her expression is no longer fearful. Now there's rage burning in her eyes that hadn't been there a moment earlier.

"How dare you?" she shouts, her voice a little hoarse.

"How dare I what?" Logan responds, seemingly unfazed by her. My eyes go back and forth between the two.

"How can you act like you have no idea what's going on? This is all your fault. *You're* the reason for everything. John made us spy on everyone, watching everything and reporting back to him so he can report to you." Beatrice pauses to take a deep breath. She begins coughing, and Mary tries to interject, but Beatrice raises her hand and shakes her head. My eyes go straight to Logan and I watch his eyebrows lift at Beatrice's words. He looks around the room at all of us listening, then turns back to Beatrice, who is nearly out of breath. "We all heard you both, talking behind closed doors, thinking nobody was listening. Well, I was listening, and I know what you're up to. Everyone in this room is going to find out." She stops talking for a moment, and I think she is going to take a seat, but doesn't. "I'm already dead, so what do I have to lose anyway?"

"Is this true?" I ask, almost too stunned to speak. We all look at Logan, waiting for an answer.

"No. The only thing we ever spoke about was the food running low. I don't care what that bitch says, she's lying."

"That's bullshit, Logan, and you know it. I heard what you said. You can't hurt me. I'll just tell them what I overheard you say." Beatrice rasps out, coughing and wiping blood from her mouth with the back of her hand.

"This is bullshit. She's obviously crazy." Logan's words land hard, causing Beatrice to lunge forward, her weak body almost collapsing. Mary rushes to her to hold her back.

"See, I told you all, she's fucking crazy!" This time, he pisses me off.

"Logan." His attention snaps to me and our eyes meet. "Why don't you explain yourself? Surely she isn't making all this up." He doesn't answer the question. Instead, he just stares, lips sealed in a straight line and brows furrowed together tightly.

It's Beatrice who breaks the silence, "Logan and John plan on—"

Suddenly, her words are cut off. She projectile-vomits blood across the floor, covering everyone's feet. We all leap backward.

"Oh, my god. Beatrice!" screams Mary, rushing to her. Beatrice continues to convulse uncontrollably, the same way Mary mentioned earlier about Jerry. The vomiting stops, and Beatrice wipes her mouth. Her face briefly clears like she is having a moment of clarity. She must realize this is her only chance to speak, so she quickly turns to me and grabs my arms.

Her face is gaunt and pale, and her eyes are bloodshot and shiny. Looking into my eyes, she utters two more words: "Killing you."

She falls backward, her eyes rolling back as she starts to seize, her body shaking violently. All her muscles tense, and her head tilts back as far as it can naturally go. White foam appears on her lips. I stand frozen, overwhelmed by the scene and her last words.

I stand motionless, too stunned to speak as Mary quickly turns Beatrice onto her side, doing everything she can to save her. The seizure drags on for what feels like an eternity. Then Beatrice's violent shaking stops, her body going limp in Mary's arms. I look at Mary and watch her eyes fill with tears. Within seconds, Mary wipes the tears streaming down her cheeks, and her expression shifts from sadness to rage. She begins breathing so hard and fast that she seems about to explode. I know exactly who is to blame—*Logan.*

Mary gently lays Beatrice across the blood and vomit-soaked floor and stands. For a moment, Mary stands there, her eyes locked on the lifeless body resting at her feet. No one dares to speak or break the silence.

"What... did... you... do...?" she says, each word slow and deliberate. Almost menacing.

The room remains silent, and Mary turns slowly, her face wet and her eyes bloodshot from crying, her jaw clenched tight. Her gaze sweeps over everyone in the room, her focus never wavering. Her eyes lock onto Logan and she stares at him as if she were a wild animal—a predator on the hunt, with Logan as her prey.

"I asked you a question," she says through gritted teeth.

"Listen, this is all a big mis—"

"What did you do?" She cuts Logan off, not giving him a chance to give her a bullshit answer.

"I was trying to tell you... This is all a big misunderstanding," Logan says, taking a step back. We all stare at him in silence.

"Is he fucking serious?" Mary asks, almost giggling like it's a joke. "How can this be a misunderstanding? People are dying and Beatrice seemed pretty fucking dead set on *you* and John being up to something." Mary's words are sharp.

The room is terrifyingly quiet.

"Answer me, goddamn it," she snaps.

"Yeah, Logan, answer her," Kelly says from across the room, her voice laced with sarcasm. "We'd all love to hear how something like this could be a... what did you call it? — *a big misunderstanding*." Before Logan can respond, someone outside the room begins screaming in pain, soon joined by two others.

"The other patients!" Mary cries out, then takes off running toward the door. "We've got to get these patients onto their sides." Everyone rushes to lend Mary a hand, except Logan, who stands there, grinning, blocking my exit and refusing to help.

"This isn't over," I say, shoving him out of the doorway. *Fucking prick. What did my father ever see in this guy?*

"I should have had John put rat poison in your food too." I stop dead in the doorway and turn to face him.

"It *was* you... I knew you couldn't be trusted. You're not going to get away with this Logan."

"I already have." A sinister grin spreads across his face.

I feel the heat building up in my face as anger consumes me. I charge at Logan with all the force in my body and he steps out of the way at the last second before I'm able to tackle him. *Fuck him and his military training*. Before he leaves, he turns his attention back to me on the floor. "No one is going to believe anything you say. I'll make sure of it. You and your little posse better watch your backs. Beatrice was right. I *am* going to kill you." I'm speechless as Logan turns away on his heel and disappears.

A moment later, I'm outside of another patient's room, only to find the same scene inside as Beatrice's room—blood splattered across the floor and the patient seizing. I scream for help, but no one comes, because they, too, have a patient facing the same fate.

Several minutes pass, and my patient dies the same way Beatrice did. I leave the room feeling physically and emotionally defeated. Mary, Julie, Tyler, and Kelly walk out of the rooms, and the sad looks on their faces tell me all I need to know. *They're all dead.*

"Beatrice was right," I say, running my fingers through my hair. "Logan just confessed, he told me that he should have had John put rat poison in my food too. So now we know what killed them."

"He can't get away with this," says Mary. "We have to tell everyone and figure out what he's up to before more people die. We can't have people killing because they want to."

"I know what he wants. He is obsessed with power, and I stand in his way. Being on this ship without anyone to tell him what to do is giving him everything he has ever wanted. He sees this as an opportunity to be fully in control. He also said that we need to watch our backs. So it looks like I'm not the only one he wants to kill."

"I swear, when I find that asshole, I'm going to kill him myself," says Mary, her tone full of fire and venom.

"I agree," I say. I look over to see the others standing in shock. Mary's breathing has intensified like she is about to explode.

"We'll find him. If it's the last thing I do, I'm going to slit his throat and take his life just like he has done to those innocent people. Logan is a monster that has to be stopped."

Everyone stares at Mary, too stunned to speak.

"Yes, we will," I say, breaking the silence. "Logan needs to be stopped for all of our sakes."

We take the bodies of the patients and dispose of them over the ledge and clean the rooms up. We do this silently, contemplating everything that has happened. I'm numb to disposing of dead bodies. I can tell the others are having a hard time processing what is happening. After a few hours of cleaning the medical center, we make our way back to our rooms. The moment we step out of the medical center into the hallway, people begin shouting.

"You won't get away with this."

"Kelly is a murderer."

"You all should be locked up for protecting her." The awful slurs continue. Someone even shouts about me being the cause of Beatrice dying. I'm confused because it's only been a few hours since Kelly killed John and Beatrice died. How did word spread so fast about Beatrice and the others?

The following day, Logan is still nowhere to be found. The ship's atmosphere has shifted drastically. More and more people approach Kelly, threatening her, telling her that her time is coming and that she'll pay for what she's done to John. Then people start threatening me, telling me they know what I did to those sick passengers. *They think I poisoned them.*

Slowly, people on the ship are becoming increasingly aggressive. At first, there were only empty threats. Then the threats became more personal, with little notes left in front of Tyler and Kelly's doors demanding that Kelly be held accountable for what she did, and then there were notes in front of my door with threats that my time was coming. When I tried to hold a meeting in the dining room to address the threats, no one would listen and people started throwing trash at me.

"Logan is coming for you," a woman's voice shouted over the crowd, from the back of the room, "and the rest of your little group." But before I could identify the speaker, the crowd burst into agreement and someone else yelled, "Why don't we take them to him ourselves?"

The crowd once again shouted in agreement. I knew my group couldn't fight everyone off. Quickly, I grabbed Julie's hand and told her to run to the medical center and lock the door behind her. Tyler, standing next to me, heard me and quickly grabbed Kelly, and gave her the same instructions. I stepped back and tried to talk my way out of the situation, but no one in the dining room would listen to me.

At that moment, I realized Logan had been slowly spreading hate to everyone, turning them against me, just like he said he would. I made a break for it, running as fast as I could. With several people coming for me in all directions, I'm nearly tackled to the ground before making it to the stairs. I make my way up to the 11th floor, hoping and praying the rest of my group made it there safely.

The door slams shut behind me when I reach the medical center. Immediately, there is pounding on the door from the other side, with multiple people shouting to be let in. Mary grabs the ropes that were used to hold the curtains back and wraps them around the handles to keep the door from opening. Tyler quickly moves heavy pieces of furniture in front of the door and others begin bringing more items to barricade us in. The shouting from the other side of the door suddenly falls silent, followed by a gentle knock.

"Help me, please," a soft woman's voice cries. "My arm is broken from getting knocked down. Please, Mary, you have to help me."

"Take down the barricade, she needs my help," says Mary, sliding a chair we just placed away from the door. I don't stop her. I walk to the side of the barricade to get closer to the door and then place my ear against it. Something is telling me this isn't right.

"Please, Mary, hurry," the woman cries again.

"I'm coming," shouts Mary back at the woman. I press my ear closer and put a finger in the other to hear more clearly.

"It's working. It's working. Lana, keep crying, she is going to open the door," whispers a male voice.

I jump back from the door. "Stop! It's a trap! They want you to open the door; I just heard them tell the woman to keep crying." Mary stops moving and stares at me.

"Are you sure?" Mary asks.

"Yes," I say, walking back to the side of the barricade. "We know you're not hurt, and we are not opening this door. Everything Logan has told you is a lie. He is the one you can't trust." A silence cuts through the air. *Did they leave?* I hear heavy footsteps approaching and a voice that chills me to the bone.

"Hello, Theo," says Logan. "All of these people know what you had John do to those poor innocent people you had spying on everyone. You can quit the hero act now."

"You know damn well you were behind all that." I fume.

"That's why you had that little cunt Kelly kill him so he couldn't tell anyone. You've been caught, Theo, so why don't you open this door and come out here and face me like a real man, you coward." His words send a burning rage through my entire body.

"With pleasure, you fucking prick." I start to undo the barricade, and Tyler grabs me.

"No!" Tyler pulls me back, and I fall to the floor.

"Let me go, Tyler," I spit.

"No, if you go out there, he will kill you and kill us. The man is a fucking psychopath."

"He is right, Theo; the door needs to stay closed," says Julie, putting her hand on my shoulder.

"I agree," says Mary, followed by Kelly. I steady my breath and just sit helplessly on the floor.

"Don't worry, you don't have to open the door now. We have as much time as it takes. Eventually, you'll have to come out. So, unless you plan on jumping through a window, you're going to need to come out of that door, and when you do, I'll be patiently waiting."

I don't say another word. We all back away from the door and sit down at a table.

"We need to make a plan," I say, gazing out the window.

19

THEO

Three weeks have passed. The constant pounding on the doors and threats of killing us has finally come to a stop. It seems that Logan's people have left us alone. Prior to today, when we attempted to open the door, someone was always waiting for us. Now, it seems that they've all left— it's blessedly quiet and we can't see any shadows under the door.

"Alright," I say, sitting at a table, staring at the makeshift barricade we placed in front of the medical center's doors. "We know what doesn't work, but now that Logan's group has moved on, we have to figure out something to ensure our safety."

"I think I have something that could work," Mary announces, her eyes glancing around at each of us.

"I'm all ears," I say.

Mary shifts in her seat and then points out multiple medical supplies, listing everything we have in stock: antibiotics, bandages, suture kits—

"I'm having trouble figuring out why you're telling us about the medical supplies that we already know are in here. Enlighten us?" I ask inquisitively.

"Think about it. We're in the only part of this ship with the supplies to save someone. I don't know why I haven't thought of this before, and I am the only one trained in medical procedures." I sit quietly, listening. "We can come up with a plan, something to make it clear or at least convince the rest of the people still on board that they need me—us," she says, quickly correcting herself.

"Okay, I'm listening. How do you suggest we do that? We've been trapped in this room for weeks, unable to leave, thanks to the brainwashed people under Logan's command. We all know Logan is responsible for this." I look around the room and Tyler, Kelly, and Julie all nod in agreement.

"I can go see what's happening out there," volunteers Tyler.

"You're not leaving me behind, no way," Kelly quickly replies, grabbing his arm.

"No one is going anywhere. Mary, how do you suggest we carry out this plan?" I ask.

"Honestly, I think Tyler has a good point. If he's willing to step out—"

"I am," says Tyler.

"Tyler, let Mary finish," demands Kelly.

"As I was saying ... if Tyler is willing to go out, I suggest we let him. He can go out in the middle of the night, make his way around the ship, and report back on what's happening." We had been hearing a lot of banging and movements, so clearly Logan's crew was up to something.

I consider what Mary is saying but I also agree with Kelly. I don't know how I feel about this. Tyler's my brother, and I don't want him getting hurt. *I have to protect him.*

"That's a big risk for my brother, and I couldn't forgive myself if anything happened to him. I agree, someone needs to check out the place, so I'll do it. Besides, I know this ship better than anyone else."

"No, Theo," Tyler snaps. "You don't always have to be the hero. Let me do this. I need to do this."

Kelly interjects, "Babe, you don't need to prove anything to me. I know you're tough, and I love you."

"This isn't about you, Kel; I love you too. But it's about me showing my brother I can handle tough situations."

"Tyler, there's no need to prove that to me or anyone else. It's not your responsibility. Stay here and protect them. Let me handle Logan. Honestly, I've been wanting to give that bastard a taste of his own medicine for quite some time," I say, smiling.

"Babe, are you sure about this?" Julie asks. "Do you really know Logan as well as you think you do?" Her question throws me off guard. She's right. I don't know him that well, and he most likely can beat my ass, but I won't go down without a fight. I don't say that out loud, of course.

"It doesn't matter how well I know him. I won't let any of you risk your lives."

"If you won't let me go alone, then I'm at least going with you. That wasn't a question, just so we're clear," Tyler states, crossing his arms across his chest.

"Fine. Tyler and I are leaving tonight. Does anyone have anything to say?"

No one says anything. Everyone helps search for anything Tyler and I can use as weapons. From what I see, there really isn't much to choose from. After about an hour of searching, I call off the search and sit in a chair, empty-handed.

"There isn't anything useful here," Mary says, holding a zipper pouch, slumping into a seat. "It's mainly medicine and supplies. Oh! Maybe this will work." She holds up a handful of sealed tools that remind me of a butter knife.

"What are those?" I ask. Mary giggles, like it's obvious what they are.

"I'm sorry, these are disposable scalpels, *very* sharp." She opens one and slides it across the table to me. The small metal blade catches the light.

"I know what a scalpel is, I just couldn't see it when you held it up."

I try to think of other supplies that may still be hidden. As soon as the thought enters my mind, I clap my hands together. "I've got it. If we can reach the atrium, I know of something we can use!"

"Well, what is it?" Tyler asks.

"On the 6th floor, there is a utility closet next to the library, and inside it is a fire hose and an axe." I run my fingers through my hair. "That's if no one has already taken it."

"Sounds good to me. But I'll need something for protection, too," says Tyler.

"For now, you can carry the scalpels; we can find other weapons along the way," I say.

"Tyler, if you're going to use a scalpel, be careful. If anyone tries to attack you, even the smallest swipe can do severe damage."

"Got it, Mary. Thanks." Tyler grabs the scalpel and accidentally drops it on the table. "My bad," he says, picking it up. *For fuck's sake, this is a terrible idea.*

Mary stands up and heads to the back of the room, disappearing into one of the rooms for a moment. When she returns, she has a black plastic trash bag in her hands. The bag is stuffed to the brim like it's Santa's bag of gifts.

"What's that?" I ask when she drops it on the floor beside us.

"Clothes I've found," she says, and I immediately pause. Hold on, what?

"You mean those are dead people's clothes? Oh, hell no. I'm not wearing dead people's clothes," says Tyler, shaking his head. I can't say I disagree with him.

Mary rolls her eyes and sighs. "Way back when we were first looking for supplies, I snagged some bags that were abandoned in case I needed material to make bandages or whatever. All that stuff's mostly clean." She waves her hand at Tyler, who is dressed in a stained white shirt and dirty khaki shorts. "You might as well be walking around with a spotlight, wearing that in the dark."

Together, Tyler and I rummage through the bag, pulling out the darkest clothing. Almost everything inside the bag is tropical wear. I pull out a button-up men's shirt with pineapples and flamingos printed on it and toss it to the side, then another with sea turtles. After tossing a few more items that are too bright, I finally pull out a dark maroon shirt and blue jeans. Tyler grabs another pair of blue jeans and a navy-blue shirt with a small ocean wave across the back and the words 'Surf's Up' printed across the back.

Several hours later, the sky is deathly black, and I stand beside the barricaded door. I'm nervous about what's to come, but I do my best not to pick at my nails. I can tell Tyler is also nervous because he hasn't said a word in the past twenty minutes, and is pacing back and forth. It's the same thing he would do when something was on his mind when we were kids. I swear, there were track marks in his bedroom on the wood floor. He was always worried about something. I gently stop him by grabbing his shoulder. He looks up and smiles. I know that nervous smile. I smile back, and then slowly and quietly, we take apart the barricade, one piece at a time.

"Ready, Tyler?" I whisper once the barricade is cleared.

"As ready as I'll ever be, now let's go before I change my mind," he says.

"And you wanted to go out alone. You know you don't have to prove to me you're a hotshot," I say, giving his shoulder a quick squeeze. "I'm your brother, I already know." He doesn't respond.

Before we leave, Kelly walks over, kisses Tyler, and gives him a tight hug. Julie is right behind her and does the same to me.

"We will be right back. I love you," I say, kissing Julie once more, and then push the door open.

The door is open just enough for me to take a peek. I don't see anyone. "All clear," I whisper. My heart is nearly pounding out of my chest. *This is a terrible idea.*

Tyler and I quietly step into the hallway. Immediately, we realize the ship is far different from how it was weeks ago.

Before Logan took over, each person had a role in maintaining the makeshift community. However, since we have been barricaded in the medical center, it appears that things have changed significantly.

In the hallway, as Tyler and I walk, I see empty cans of food and empty bottles of water everywhere. It looks as if the place had been ransacked. The smell of urine lingers in the air, suggesting the man-made bathrooms are no longer in use or that people no longer care and piss on the floor. We step lightly, making our way to the stairwell. Each of us is doing our best in the dark not to fall. The trashed hallway isn't making it easy to maneuver or stay quiet. Once we reach the stairwell, we begin to descend carefully, watching each other's backs.

Standing on the 6th floor, we find the closet that's supposed to have an axe. I slowly open the door. *Please be there, please be there, please be there.* The hinge squeaks, and I stop moving. I hesitate for a second, looking around to make sure no one is around. In one quick motion, I pull the door open fast so it doesn't creak. A smile spreads across my face as I notice that not only is the axe still in place, but there are two. I let out a sigh of relief.

"Tyler, there are two axes, so we can each have one," I whisper excitedly.

"Thank god," whispers Tyler.

After grabbing them, we turn and face each other. 'Ready?' I mouth, and he nods back, giving me a thumbs up. My heart is pounding again, and I feel beads of sweat forming on my forehead. I look over at Tyler, his hands are gripping the axe so tightly that his knuckles are turning white.

Slowly, I crouch down and walk toward the railing directly in front of the library entrance. Tyler follows my lead, crouching down close behind me. We stop at the railing and I set my axe across my knees. Looking through the railing, we have a bird's-eye view of the atrium. My senses are on high alert, and I do my best to stay quiet. I go to set my axe down on the floor and reposition myself, but then stop instantly when I hear a noise from down below.

"What is it?" Tyler whispers.

I put my index finger to my lips then point two fingers at my eyes, then point them toward the railing to get Tyler to take a peek. As soon as Tyler gets a view, his eyes widen, and he leans back away from the railing.

Standing directly below us in the atrium are two men whispering. I can't see their faces or understand what they are saying but they seem to be arguing. Immediately, I noticed how much the atrium has changed since a few weeks ago. All the entrances have been securely blocked off with broken furniture and tables. From above, I point out to Tyler that there is only one way to enter the atrium, and that's the stairwell behind the elevators. While I'm doing so, I spot Logan walking in. A knot forms in my stomach and I feel heat rise in my cheeks at the sight of him. The two men who were whispering have now gone silent and are standing straight up, facing Logan like soldiers.

"What are you two nitwits arguing about this time?" Logan asks.

"Nothing, sir. We were just talking," says the man on the left.

"Did you do what I told you to do, Randy?"

"Yes, sir, we did," answers the other man.

"Well...?" Logan asks, sounding irritated.

"We didn't find anything; all the water is gone. They took everything, and the back of the ship is completely blocked. If we try to enter, I'm sure they'll kill us. They don't like you or any of us who work with you," Randy says, his voice soft like that of a child that's in trouble.

"I guess we'll have to find a way, and if that means killing everyone there, so be it. I'm no longer negotiating with them. They'll either give us the water, or we attack, that's an order." Logan pauses, looking at Randy. The other guy doesn't move or speak. "We have all the food on the ship, so unless they want to starve, they'll agree to our demands."

"Yes, sir, but uh... how are we going to do that? We don't have enough weapons to attack with."

"*They* don't know that, dumbass. We'll give them one more opportunity to provide what we've asked for and issue a warning. Their so-called leader, Chris, is a fucking joke. He won't call our bluff."

"But what if he does?" asks the other man nervously.

"Then I suppose they'll kill *you*, Todd. Unless you want that to happen, I recommend you convince them you're not bluffing. Now, get the fuck out of my face. Gather the others and meet me back inside the Stardust Lounge at sunrise so I can share the plan with everyone," Logan demands, then walks back out of the room.

Tyler and I remain still as they all exit the atrium. Once the room is clear, I look at Tyler. "We need to tell the group what we just heard."

"Agreed," says Tyler.

When I start to stand, the axe slides off my lap and it lands with a loud thunk. Tyler has already begun walking and turns when he hears the sound. Before I can bend down and pick it back up, Roger– an overweight man that closely resembles an ogre, appears from around the corner. I recognize him from working with John.

"Tyler, behind you." I snap, Tyler turns around, and then jumps back toward me, away from Roger, who is now staring at us from a few feet away. His big belly is peeking out from under his shirt, his beard is overgrown and stringy, and he is holding a wooden stick with one end shaved to a point. He takes a step closer and I get a whiff of his rancid body odor that nearly gags me. He smells like he hasn't tried to bathe in weeks.

"Look what we have here... Logan's going to be real excited to see you," says Roger, his eyes locked on me.

I contemplate how to handle the situation. I shift my head and glance down, realizing I never picked up my axe after it fell. "Logan doesn't need to know," I say, keeping my voice calm. "We'll just go back to where we were and leave you be. There is no need for things to get ugly."

"Things are already ugly, Theo, and I'm not letting either of you go. There's a price on your head and I want to be paid."

"How much?" asks Tyler.

"Five cans of beans for Theo, and you, my friend, are worth two. Looks like I'll be eating for a while," he says, licking his lips.

"Well, that's just fucking rude. You look like you can go a few days without a meal," replies Tyler.

Roger's grin quickly turns into a snarl, and his face turns red. Tyler has pissed him off, big surprise there.

"Shut up, Tyler," I say through gritted teeth. "Come on, I can give you food if you let us go." *We need to do something, fast.*

While I'm trying to negotiate our way out of this situation, I feel Tyler subtly bump into me. I let out a cough and glance down. Tyler is holding his axe behind his back. I can't bend down to grab mine, so I shift my body slightly to get closer to him and put one hand behind me. Immediately, Tyler understands what I'm doing.

"No, there'd be a price on my head if I did that, and I can't have that. Don't think about moving a muscle, or I'll take this and find out what your insides look like," says Roger, lifting the broken broomstick. *Things are so much worse than I thought. We are past working together. These people are going to kill us all.*

Roger turns his head to call for back-up, Tyler quickly puts the axe in my hand and I swing it, aiming directly for the back of Roger's head.

"Hey, Carmen, come ove—" Roger goes silent mid-sentence and slowly turns around to face me. His face is plastered with shock, his mouth wide open. Silently, he lifts his hands to his head, frantically feeling around. His fingers are painted crimson as they touch the wound.

"Now we can see what your insides look like, you asshole," I say, no longer whispering.

Roger stands there, paralyzed, staring at his bloodied fingertips. Blood begins to drip down his forehead and into his mouth. Within seconds, his legs give out, and he drops to his knees, falling forward and face-planting the floor. The large axe is embedded deep in the top of his skull. I walk over and place my foot on the back of the man's head. *I can't believe I just did that. There is no turning back now.*

I look back at Tyler, his mouth open in shock, staring at me. With two hands, I grip the handle of the axe and pull. The axe flies out, and blood splatters across the floor and onto my face. I look down and stare at the blood pooling under Roger's skull. *I just killed a man. If I didn't, it would be me lying in a puddle of blood.*

"Holy shit, Theo!" Tyler says, pulling me out of my thoughts.

I feel Roger's warm blood dripping down my face. I'm frozen as I process what I just did. His blood is slowly inching toward my mouth, and I instantly reach up, using the back of my hand to wipe it off. *Things will never be the same from here on out.*

"It's survival of the fittest now, Tyler." I look down at the blood-soaked axe in my hand and hold it up for Tyler to see. "Don't be afraid to use your axe. Let's find whoever the fuck Carmen is and kill her, too. I'm done with this bullshit."

Tyler doesn't say anything. He just follows my direction. We step over Roger's body and head toward the stairs. As we approach the corner, I raise my hand to stop Tyler from moving forward. I watch as someone I assume is Carmen makes her way from the other side of the ship, also carrying a makeshift weapon —one like Roger had in his hand. I quickly estimate the distance and look at Tyler. "We have about fifteen seconds before *Carmen* gets here."

"What are we going to do?" asks Tyler, his voice nearly cracking.

"This time, you need to be the one to swing the axe."

"What! No, I can't do that. Theo, I won't do that."

"Tyler, if you don't do this, she'll kill you. Everything has changed. You heard what Roger just told us. Logan wants us. They are going to kill us. Just swing as hard as you can into the hallway, that's all you have to do."

"Fuck. Fuck. Fuck," says Tyler, taking a deep breath. "Okay, Theo, I can do this. Tell me when."

Tyler's hands are gripping the axe like a baseball bat. His fingers are so tight that his knuckles are white. I crouch down and peek around the corner. "Okay, Tyler, get ready," I whisper. I watch Carmen approach. She is looking down at the floor, not paying any attention. "Three... two... one, now!" The sound of my voice makes her look up.

The axe's edge strikes directly into Carmen's throat. Blood instantly gushes out around the blade. The axe hits her so quickly that she has no time to react, and her expression remains the same. Her hands fly to her throat and grip the axe that's still embedded. Tyler pauses for a split second, stunned. She tries to speak, but only the sound of gurgling echoes through the hallway. "Tyler, pull the axe out!" My voice snaps him out of his trance. He glances at me, then back to Carmen. His right hand is still gripping the axe's handle. He lifts a foot and kicks Carmen backward, the axe dislodges, and blood sprays out, splattering several feet. Her body twitches as she slams to the floor with a loud thump. I stand up and take the axe from Tyler's shaky hand. Tyler doesn't move his face or body, trembling in shock. I grab his shoulder, pulling his attention back to me.

"Tyler, come on, we need to go," I say, pulling him toward the stairwell. "Tyler! We have to go before others arrive."

Tyler looks up, and our eyes meet. His expression is hollow, like his soul has just been ripped out. The blood splattered across his face makes the whites of his eyes stand out and he's staring as if he doesn't understand what I am saying. I pull his arm again and he begins walking, following silently like a ghost behind me.

20

THEO

The door to the medical center is locked. I pound my fist against it. "Let us in!"

After a moment, I hear shuffling behind the door, then the door flings open. Mary steps back from the doorway, letting Tyler and me step inside.

"Oh my god, what happened?" Kelly gasps, running up to Tyler, who still has the same shocked look on his bloodied face.

I began to speak, but Julie interjects, "What happened to you? Why is there blood all over both of you?" she asks, running her hands over my chest and arms, looking for an injury.

"We're okay," I say, looking back at Tyler. "Well, I'm fine. Tyler's in shock. We both killed someone."

"Killed someone?" Mary asks, quickly closing the door and restoring the barricade.

"Tell us what happened," Julie demands.

"Can you find me something to clean myself up with first?" I say, wanting to get the blood off my face.

"I'll grab you something," says Mary.

She runs to the closet and brings a box of antiseptic wipes

over to Tyler and me. I take one and do my best to clean my face. Kelly helps Tyler while he stares blankly at her.

"I... killed someone," his voice trembling, almost inaudible.

"It's okay, babe. You're safe now," Kelly says, rubbing his back in a soothing motion.

"The situation is far worse than we thought," I say, still scrubbing my face.

"What do you mean?" Mary asks, opening another pack of wipes to hand over.

"He means we're in a fucked-up situation, and I don't know how we'll get out of it," Tyler says, quietly.

We all quickly turn to look at Tyler. I'm surprised he's speaking. Then his eyes shoot to mine and his entire mood shifts. There is an anger in his eyes I've never seen before.

"Go ahead, Theo, tell them," he pauses, staring at me. "Explain what you did, and what you made me do." Everyone's gaze lands on me.

"I only forced you to kill that woman because if you hadn't, she would have killed you. They all want to kill us!" I snap. Turning away from Tyler, I run my fingers through my hair and let out a long breath.

"Ok, look, can you just tell us what happened?" Julie demands.

"It's bad, but we did find Logan. He is the one in charge—actually, he's only in charge of half the ship."

"Half the ship? What do you mean by that?" Mary asks.

"I'm not entirely sure. When Tyler and I made it to the atrium, we overheard two of Logan's men, Randy and Todd, arguing. We couldn't hear what they were saying, but Logan walked in, and we heard him ask about something he'd told them to do. They said they had, but they couldn't find anything, then—"

"Find what?" Julie asked, cutting me off.

"I'm getting to it. From what I understand, Logan and his group have all the food on the ship, and some guy named Chris has all the water."

"I know Chris," says Kelly, her face brightening. "Well, I know of a Chris. Not sure if it's the same one."

"Who is he?" I ask.

"Chris is Sara's boyfriend. I haven't seen either of them since—" Kelly trails off, looking down at her feet.

"Since you killed John?" answers Julie.

"Julie!" I shout.

"What? She did kill him and now you and Tyler have killed someone too." Her words linger in the air for a moment.

"Hold on a second, calm down," I say before continuing. "Apparently, this Chris guy controls all the water and he refuses to share it with Logan. But here's the catch: Logan, supposedly, is working on a plan. He intends to send his group to the other side of the ship and demand water, threatening to kill them if they don't give him what he wants."

"So what does that mean? What's your plan?" asks Mary.

"Now that we think Kelly's friends are on the other side, we'll go there and explain what Logan has planned," I answer.

"Okay, I like that idea. But what if it's not the same Chris that Kelly believes it is?" Mary makes an excellent point. I ponder her question for a moment.

"You are right; it may not be. But either way, we know about Logan's plan and we still need to tell them."

"Good point." Mary nods. "Now, can you explain why you are both covered in blood and whatever it is Tyler said you made him do?"

"Yeah, about that — remember Roger, the guy that John had working in the kitchen with him, a bigger guy, kind of intimidating?"

"Yes..." answers Mary.

"We ran into him when we tried to leave from the 6th floor. He appeared out of nowhere and cornered us. He told Tyler and me that we have a bounty on our heads." I trail off for a split second. "Now that I'm thinking about it, I wouldn't be surprised if all of us had a bounty out on us. Logan wants us captured. And the bounty is worth… wait for it — five cans of beans."

"No, Theo, *you* are worth five cans of beans. I'm only worth two, which means he is really gunning for you," says Tyler.

"Is food really that scarce out there now?" Kelly asks.

"Seems that way," I say. We all stand silent for a moment, processing what this means for everyone.

We devise a plan to locate Chris and his group and figure out how to make contact without being seen by Logan's group. Since Tyler and I have already left once, everyone decided we would be the ones to go out again and find Chris. I wasn't sure if Tyler would be able to handle going out again, with what happened last time. But he didn't refuse, and I didn't want to bring it up. While Tyler and I were out, Mary, Kelly, and Julie would gather as many medical supplies as possible in case we all needed to evacuate if Logan attacked. Mary suggested that once Tyler and I find Chris and his group, we offer to have them relocate to the medical center, since it's one of the largest secure areas of the ship, with only one way in and one way out, unless you scale the side of the ship to get in from the balcony. Luckily, no one has tried that yet. Being on the 11th floor, you would fall to your death if you slipped.

After cleaning up and sleeping a few hours, I help pack up the medical center. Mary disappears to the back for a few minutes, then returns with a large duffel in hand. She pulls everything out to inventory, and it's packed with essential medical supplies, including surgical scissors, saline bags for IVs, suture kits, and burn creams. Julie grabs a bag and starts grabbing anything that may be useful.

"What can I do to help?" I ask, empty-handed.

"Follow me," Mary says, getting up and walking toward a storage closet where Kelly is standing. Inside the closet, Mary has shelves full of powdered baby formula in cans.

"Do we need to pack this? Where the hell did it even come from?" I ask, a little confused.

"Yeah, pack it. It could come in handy if we run out of food, and it came from the daycare on board. There was an entire stock of it," she tells me.

"Holy shit, I forgot about that!" The board of directors insisted there be adequate supplies for the childcare room for the future sailing with families. *Not sure why that was a priority, since there weren't kids on this sailing, but unfinished floors were easily overlooked, but what the hell do I know?* I don't say that out loud. Instead, I help Kelly pack the formula in the bag.

"Well, I am glad that they did because I used the formula for one patient right after the explosion. He couldn't eat solid food at the time, and believe it or not, the baby formula provided enough nutrients to keep him stable," explains Mary.

"Huh, I never thought about that. I guess it would be like a protein shake for an adult." I finish helping Kelly, then walk back out to find Tyler gathering blankets, towels, and sheets, and placing them in a cardboard box from a storage closet. The sight reminds me of the secret that's slowly eating me alive with guilt. I make a mental note to keep track of this box because winter will be coming very soon and we will need all the clothing and blankets we can find.

The sun has set and the night has fallen. Tyler and I are

getting ready to leave to go hunt for Chris. Mary once again removes the barricade to let us exit. As we are walking out, Mary grabs me on the shoulder. I stop and face her.

"Good luck." I nod, not sure what else to say. I turn back around to Tyler. The door shuts behind us and instantly I hear the shuffling sounds of the barricade being put back in place.

Quietly, we creep through hallways and stairwells, descending to the lower decks, checking each floor for movement or noise. On the 1st floor, there are only crew areas, including the engine room, and there is no sign of anyone else. Before we leave, I notice another service closet, and I find two more axes inside. Tyler and I each grab one and tuck the other into our belts. As we are leaving the 1st floor, I remember that Logan mentioned Chris's group being aft of the ship.

"Remember, Logan said Chris's group is at the back of the ship, let's head that way." Tyler follows me. The 2nd floor is unfinished, but we check anyway before continuing upward. By the 6th floor, we are running out of potential locations. I do my best to remember the lounge locations and recall that the 6th floor is where the two biggest ones were. I knew that the Stardust lounge was occupied by Logan, so that left the Paradise Cove lounge at the opposite end. Unlike Stardust, Paradise Cove lacked an outside patio. It was smaller, more secluded, with a bar, dance floor, and booths.

As we make our way toward Paradise Cove, I begin overhearing voices—not just a few individuals, but a lively crowd. It's a clear sign that maybe we've found Chris and his group. Quietly, Tyler and I head toward the entrance of the lounge. Before reaching it, we pause, standing about twenty-five feet away to observe our surroundings. Two guards stand at the entrance. They're deep in conversation, both oblivious to us sneaking up on them. As Tyler and I start to head over to make contact, someone suddenly ambushes us from behind. They must be men because the arm that has me in a chokehold is hairy and muscular.

"We're looking for Chris. We're not here to fight!" I shout, hoping that they realize we are not with Logan's group. *I hope they're not with Logan's group.*

Seconds later, the attackers release us, and Tyler and I are both pushed to the ground. I was right — the attackers are men.

"We know who you are. Why do you want to see Chris?" asks one of the men. His face is familiar, but I can't put a name to it.

"We want to join Chris's group," I say, quickly rising to my feet. "We have supplies and crucial information he'll want to hear."

"You can just tell us and we'll let our leader know. How can we be sure you're not with Logan's camp?"

"It's about Logan, but I'm not telling anyone except Chris. He's the one in charge here, right?" I ask.

Tyler is on the ground, an arm's length away, with one of the guards standing above him. There is an audible clicking sound when the doors to Paradise Cove open. A tall woman with long brown hair steps out. I can't make out her face until she steps closer and the two guards step aside.

"Actually, Chris isn't the leader of this group; I just allow Logan to believe he is. I'm the leader here. What information do you have?" She walks towards us. When she gets closer to us and is able to make out who we are, she lets out a sigh of relief. "Theo, Tyler. It's nice to see you two."

"Sara!" Tyler jumps up and holds his arms out wide.

"Let them go, they're not with Logan."

"Man, I'm so glad to see you. Kelly has been worried sick about you," Tyler says, offering Sara a hug. She smiles and accepts.

"Oh God, I'm glad she's okay. Where is she? And what do

you all know about Logan? That narcissistic asshole started killing people who wouldn't listen to him right after that girl died in the med center. At first, the rumors blamed John for those deaths, then claimed Logan and John were working together. Then Logan went around telling everyone it was you. Anyone who questioned him was killed. Logan is fucking bat shit cra– I'm sorry I don't like to use that word."

"What word?" Tyler asks.

"Crazy," she whispers, going silent for a moment, lost in thought. "That word hurts people." Then her eyes snap back to me. "Chris and I had a feeling that Logan and John were up to something, because before that stuff with Soph– uh, Kelly." Sara pauses, then she runs her fingers through her hair, letting out a deep breath.

"Anyway," she continues, "we kept seeing the two of them whispering when they thought no one was watching. So when the shit started hitting the fan that day, Chris and I grabbed all the water and whatever food we could muster up, and we rushed to anywhere we could find to hide. That's how we ended up here in this little hideout. We heard you're all in the medical center with that nurse, and no one has been able to get in."

I study Sara for a moment before speaking. I can't stop thinking about that little hiccup with the name a second ago. Not wanting to make her suspicious of trusting me, I nod and go on like it hadn't happened.

"Yeah, that's true." I glance over at Tyler, who isn't saying much, then back to Sara. "Logan killed that girl, Beatrice, and her friends."

"How do you know?" Sara asks, her eyes locked on mine.

"He told me." Her eyes go wide in surprise.

"What? He just came out and said, 'I killed them'?"

"Without so many words, yes. Said he should have put rat

poison in my food too." I look around, getting nervous because we are standing out in the open. "Can we come in?"

"Yeah, sorry, come inside before anyone else sees us."

Once inside, the guards quickly shut the doors and put their makeshift barricade back in place. It's similar to ours with a bunch of random things. My eyes widen when I see a large group of people silently staring at Tyler and me.

"Holy shit," Tyler murmurs beside me.

"We have seventy-eight people in here, so it's a little cramped; this is the only place we could all get to without worrying about Logan and his group getting inside," explains Sara as Chris walks up and stands beside her.

"Thank you for trusting us and letting us in," I say, extending my hand out to shake hers. She doesn't take it; instead, she looks at Chris, then back to me.

"You're welcome, but I never said I trusted you." She lets her words hang in the air while she stares at me. I don't protest. I look past her and get a good look at the area.

The room is dark, with only a faint glow in the distance from a small flashlight in a metal bowl illuminating the space. Paradise Cove is at the back of the ship and offers very little natural light from outside. There are a few small porthole windows in the back of the room that only let in daylight. This lounge was originally created to be dark and intimate, like an old piano bar. The air here is stale and musty, with no airflow. The smell of urine and feces lingers; it's clear these people have been confined in this room without water to wash themselves.

Before the ship was divided, there were designated areas for people to use the bathroom and wash up. Along with those areas, there were assigned jobs for gathering water from the ocean for washing, as well as cleaning the buckets used as toilets. With no access to the outdoors, it's evident they can't do that here. Suddenly, I'm grateful that Mary was smart enough to have her crew gather multiple containers of saltwater for her patients to give them sponge baths. If she hadn't, I assume this is what we would look and smell like. Sara starts walking towards the center of the room, and Tyler and I follow close behind.

Even though the room is dark and I can't make out everyone's features, I can feel all their eyes on me.

"Why are they here?" a voice booms from the back of the room.

"Yeah, why are they here? They're not one of us. How can we trust them?" another person shouts. Like a domino effect, one by one, people begin demanding to know why Tyler and I are inside their safe space. I keep quiet, not wanting to overstep by answering. Instead, I wait for Sara to answer.

"Quiet, quiet. These are our people. We are going to help them, so let's welcome them. You all know Theo and Tyler, and I can assure you they are not with Logan. With their help, we can stop Logan and his people," Sara says, addressing the entire crowd.

From the back of the room, a man speaks up, "And how exactly are they going to help us?"

"We can help because we know things you don't," I answer.

"And what exactly is that?" asks the man. I stand silent for a moment, unsure if I should announce what we know. I look over at Tyler and he is silently staring at me, waiting to hear what I say.

"Yeah, what information do you have?" asks Sara, cutting in.

"Before I tell you what we know, I need to make sure you are not going to turn on us and force us out," I say, turning back to Sara. "Can we trust you?" Sara purses her lips together, then turns to the crowd. She doesn't answer my question.

"We're all family here now. We all want to survive and get

back home; we can't do that unless we work together," she explains.

"I can appreciate that. If you decide to trust us, we should move all of you to the medical center. There, we have fresh air and natural light from the patio, as well as more room than inside here. It's not massive, but it will give you all a little more space."

"Why don't we just kill Logan so we can all go back to our own rooms instead of being cramped together and scared for our lives?" asks a woman from the back of the room.

"Natalie, we've already discussed this," Sara says, shaking her head. "We're not just going to go out and start killing. We are not like Logan." Her voice is stern, as if she has had to repeat this multiple times.

"Has anyone tried talking to Logan?" I ask, curious if they have even attempted to have conversations.

"Yes, one girl, Brittney, was trying to make peace with the other side. We all warned her that Logan didn't care; from what we saw, all he wanted was power and to be in control of everyone, but she decided to go anyway. The following night, they brought her body and tossed her in front of the guards on duty."

"Wow, that's really fucked up," I say.

"Yeah. Ever since then, no one has left. We only have enough food to last another day or two if we ration carefully. Our water is also starting to run out." She sighs.

"We're running low too— we don't have enough to feed this many people. But that's not our biggest worry right now. We overheard Logan, he's planning an attack against you. If you don't provide them with what they want, they'll start killing—"

Everyone gasps, and panic ensues before I can finish my sentence.

"Wait, hold on," I say, raising my hands to get the crowd to quiet. "The good news is, other than the few spears they've created, they don't have any real weapons. We also heard them say that."

"Well, we don't have any weapons either," says Sara.

"Where did you get those axes? Could we get more?" asks Chris.

"There should be one or two above each fire hose station. They're located all around the ship. I can't remember exactly where each one is, though," I answer.

"Well, that's good. We just need to find them," says Chris.

"That's easier said than done. I think we should relocate to the medical center. Everyone should find something to protect themselves from Logan's group. It's pretty much a straight shot back there. We can create a plan from there on what to do next. The only thing I know for certain is that we are now in a kill-or-be-killed situation," I say.

"Okay..." Sara nods. "Give me a minute to think about all of this."

The room is silent as Sara paces back and forth. Before Sara can say no to her group moving to a new location, I decide to speak up.

"I have no idea when Logan plans to strike, so the faster we relocate, the safer we'll be. If he finds out that you have no weapons, they'll just break down your barricade and attack regardless."

"You're right. We move tonight." Sara looks around at her people. "I know that I have said countless times that we are not going to just start killing people, and I meant that. But if Logan's coming to attack us... don't go down without a fight. Theo, we'll follow you." I nod. "Everyone, grab anything and everything you can to protect yourself. Look around, these are your allies, your family. We need to protect one another. Don't let your guard down, stay on high alert, and most of all, let's get there as quickly and safely as possible."

21

THEO

The room springs into action. Everyone frantically begins searching for something to defend themselves with. I glance around and notice there isn't much in this lounge that could serve as protection. Behind the bar, a few people cheer when they find corkscrews, and some people grab empty liquor bottles from the trash, wrap them in bar towels and linens, and slam them to the ground. When they remove the cloth, all that's left of the bottles are jagged edges, ready to be used as weapons. A few also manage to find a box of unopened paring knives hidden in a drawer.

One by one, everyone in the group gathers their water and any remaining canned food and brings them to the center of the room. Once everyone brings their items over, silence almost entirely takes over, broken only by heavy breathing, as everyone waits patiently for Sara to speak. Sara makes her way to the center of the room, placing the flashlight on a table and aiming it toward the ceiling to cast a little more light across the room. She looks around, then begins to speak.

"I know we are all tired and scared, but the medical center has more space and fresh air." She pauses, looking then looks around the room until her eyes land on me. "Theo, can you please explain the plan?" All at once, I feel everyone's eyes on me while I walk over to Sara.

"Thank you, Sara. The medical center is five floors above us —We just need to head up the stairs. Assume they are waiting for us, and be ready to fight. Form a single-file line and walk quietly." I turn to face Sara. "If you think it's a good idea, one of your people and I can go up to make sure that the medical center is ready. With us moving this many people, I don't want to get there and not be able to get in. We'd be sitting ducks for Logan to attack. Sara, you want to go with me?" She doesn't respond right away.

"No, I'll stay here with everyone and make sure we have everything we can bring. Take Chris with you, and leave Tyler with me."

I look over to Tyler. "Are you okay with that?" He doesn't answer me, just nods in agreement. I shift my attention back to Sara.

"Chris is going to need protection and must be able to keep up with me. Can he handle that?" I ask, purposely looking at Sara to let her answer. She looks over to Chris.

"Yeah, I can keep up with you, and I agree, I do need a weapon," Chris answers.

"Here, use this then," says Tyler, walking up and laying the extra axe in his hand.

Chris and I make our way to the door. The barricade is removed, and the door is opened. Before we leave, I turn so that everyone is looking at me.

"We will be back soon. Be ready to leave when we return. We won't be gone long, and remember what I said. This is now a kill-or-be-killed situation. That means if Logan and his group show up, don't hold back. If they show up, it's to kill you. So fight like your life depends on it," I pause, looking around the room at the many faces staring back at me, "because it does. Chris and I will move as fast as we can. See you in a bit."

Outside the lounge, the hallway is dimly lit by moonlight streaming through the windows that line it. The air is stagnant and stale, and I catch a faint whiff of mildew, reminding me of the smell inside the library. Chris and I are careful not to make any sudden movements or sounds that could alert anyone to our presence. I occasionally turn back to make sure that he is following closely. He must be terrified. His jaw is clenched tight and he keeps looking back and forth around the area, a bit frantic. Beads of sweat have formed on his forehead and upper lip. His breathing is rapid, almost as if he is on the brink of a panic attack. I stop walking and turn to face him. Gently, I lay my hand on his shoulder, take a deep breath, and exhale slowly. After a few tries, he finally mimics me and begins to calm down. Once he is calm, we start walking again. As we approach the entrance to the stairwell, I hear a voice, causing me to stop in my tracks.

"Did you hear that?" I whisper. Chris's eyes are wide and he is staring at me.

"Hear what?" I quickly turn around and shove Chris against the wall, my arm pressed against his chest and my hand covering his mouth. Slowly, I lift a finger over my lips to tell him to keep quiet. I feel his heart begin to pound rapidly against my forearm, and the rapid breathing starts again. I do the slow and steady breaths, making him mimic me again for the second time in the past few minutes. Once he's calmed again, I lower my arm.

"Someone is coming down the stairs. I can hear them."

Chris nods but stays silent. After a moment, the voices grow louder and clearer, sounding like a woman and a man. Chris starts to tremble as sweat drips down his face as he panics. I grab his face and make him look me in the eye.

"You've got this, okay? Just be ready to swing with your axe—grip it tightly—in case someone notices us. And don't hit me with it by mistake, okay?" Chris stays silent and just nods. I look down and see his hands shaking like a leaf.

"Don't move and hopefully they won't notice us."

The two voices are so close now, they sound like they're practically beside us. Chris is so nervous and shaky that the axe he is holding taps against the wall, making the slightest noise. Oh fuck, I hope they didn't hear it. The voices go silent and their footsteps stop. My heart begins to pound so hard I can feel it in my ears.

"What was that?" the woman says.

"What was what?" the man asks, and I hear some shuffling sounds. "I didn't hear anything. Let's go, I'm sure it's nothing. Everyone is too scared to come out of hiding. They all know Logan is nuts."

"Yeah, you're probably right," the woman replies.

"If we do see anyone, then we kill them. Or he will kill us."

Kill or be killed.

Standing with my back against the wall, I quietly exhale. I look over at Chris, pointing at my eyes, then at him, gesturing for him to keep watch and stay put. His eyes go wide and he shakes his head quickly and raises a hand. I ignore him and slowly begin to make my way into the entrance of the stairwell. I stop when the woman speaks again.

"Brent, I really think someone is here. I know I heard something," says the woman.

"Karissa, there are noises all over the place. The ship is rocking. It could have been anything. But if it makes you feel better, I'll go look. And if there is someone coming for us, I'll knock them out. How does that sound?"

"Thank you," she says. They stop talking and I can hear the footsteps getting closer once more. Slower this time. I look back at Chris and he is clinching his axe so tightly that his knuckles have turned white. He stares helplessly at me. I lift my axe above my head with the blunt end down. Each step that gets closer, I feel my heart pound harder. *Just turn around, don't make me hit you. Just turn around, don't make me hit you.* The steps don't stop. I can hear Brent breathing. He is so close. As soon as he walks past the wall we are standing behind, I slam the axe handle down on his head. He drops instantly to the floor–knocked out, or dead, I'm not sure. I step out into view and stare at Karissa. She stares back at me, unmoving, with a look of fear in her eyes. She is just a few feet away.

"Shhh," I say, pointing the axe in her direction. "Don't scream. If you scream, I will kill you. I heard what your friend just said– don't make me kill you."

I watch her eyes drift down to Brent on the floor, then back at me. A split second later, she takes off down the hallway. I drop my axe and take off after her. Within seconds, I close the gap between us. Just as she lets out a scream that's loud enough to wake the entire ship, I reach out and grab her ponytail, yanking her head back. Her feet fly out from under her and she slams backward onto the floor. I quickly jump on top of her and pin her to the floor, covering her mouth, hoping her scream didn't alert anyone. I glance back at Chris, and he suddenly picks up my axe and runs it over to me.

"What can I do?" he says breathlessly.

"Watch him, and if he wakes up, knock him out again. He might already be dead, though," I say. Karissa tries to scream under my hands at those words.

"Okay, okay," Chris says, running over to Brent, and I bring my attention back to Karissa squirming to escape.

"I said not to scream or I'll kill you. Don't make me kill you." I take a deep breath and sigh. "Again, I don't want to hurt you, but if you don't listen and do what I say, Logan will be the least of your worries. I will take this axe and slice your fucking throat open. Do you understand?"

The woman goes still, then nods. "Okay, good. When I lift my hand, I don't want to hear you make any noise," I say as I slowly move my hand.

"Hel—" she starts to scream, and I slam my hand back down to cover her mouth.

"Damnit. I told you not to do that. Do you want to die? Is that why you won't fucking listen? I'm not the bad guy here. If you would stop screaming for a goddamn second, I will explain myself." I watch tears roll down her face, and a pang of sadness runs through me.

"Don't cry. I said I didn't want to hurt you. You're the one making things difficult. So, one more time, I'm going to lift my hand and we can talk. If you scream, I'm not giving you another chance. I'll jab this axe into your throat so fast you won't even know it's coming." I slowly lift my hand again. This time, she doesn't scream.

"What do you want?" she asks, and I can hear the fear in her voice. I can't blame her; I do have her pinned to the ground with an axe in her face.

"I want to know everything you know," I say pointedly.

"Like what? I don't know anything."

"Don't fucking lie to me. Where is Logan and what does he have planned?" I demand through clenched teeth as I push the axe tighter to her throat.

"Okay, okay... I'm not exactly sure, but he said we all needed to get ready because he has a big plan for us tomorrow night, but that's all I know. I promise."

"Why are you even following him or listening to anything he says if he is going around killing everyone?"

"Because I don't want to be killed, that's why," she says, her voice sharp.

"I don't want to hurt you, and I meant that. But I can't let you go back to Logan, so you're coming with me," I say.

"Why?"

"To help us take him out so he doesn't kill anyone else." I point to Brent on the floor. "Your friend over there can come too, if that makes you feel better."

"I'll come with you and tell you what you want to know, as long as you promise to keep me safe. We also need to find my brother," she says.

"I'll do my best to keep you safe. Our people are not going to hurt you. As for your brother, if he wants to get away from Logan and join us, we'll protect him, too."

"That's enough for me. And that guy is an asshole, so you can just leave him to die. He does everything Logan tells him to. He is like a little puppet."

"Does Logan know who your brother is? I ask because I don't want Logan to see you're missing and then kill your brother, thinking you're a traitor," I say.

"No, I don't think so. We stay quiet and just do what he says to avoid drawing any attention."

"Okay, let's get going." I have a ton of questions for her when we get back to the medical center.

"Where are we going?" Karissa asks.

"You will see. It's a safe space," I answer, turning to walk toward Chris. Brent is lying on the floor and starts to groan. "We need to move fast. He is going to be waking up soon." As the three of us climb the stairs, I stop to face Karissa. "Should we expect anyone else to surprise us?"

"No, it was just me and Brent on patrol, we were looking for anyone on your side," she says, pointing at Chris.

"Logan's going to kill you and anyone who stands in his way of power. There is something wrong in his head, and I think he is slowly losing his mind," says Karissa. Chris sucks in a deep breath at her words.

"I just had an idea. Did you really mean it when you said that you want to take Logan down too?" I ask.

"Yes, he has power over everyone, and whatever he says, people just follow him out of fear that he will kill them. Some of us are living in constant fear."

"Okay. Well, if you want to, it's going to be tough, and you're going to get hurt." Karissa doesn't answer right away. She just stares at me.

"What do you have in mind?"

"Yeah, Theo, what's your plan? I'm ready to get out of these dark hallways," says Chris.

I look around to make sure no one else is in the hallway listening. "I'm going to need to punch you in the face."

22

THEO

The medical center is quiet and dark as we approach its doors. I gently tap my fingers on the door three times, looking around to make sure no one is sneaking up on us. A few seconds pass, and no one answers. I lightly tap three times once more and wait.

"Why are they not answering?" Chris asks impatiently.

"They will. Let's give them another second," I respond.

Several seconds pass, and I'm about to start pounding on the door a little louder when I hear shuffling behind it. The door slowly opens a crack. It's Julie; she hesitates a second when she sees Chris and Karissa standing behind me.

"They're with me," I say, pushing the door open.

"What happened? Where is Tyler?" Julie demands.

"Tyler is fine. Let's close these doors so I can explain everything."

Once we are all settled inside and the door is re-locked, I ask everyone to join me at a table. The room is quiet, and the balcony door is open, letting in a cool, refreshing breeze. Kelly quickly zeros in on Chris and hugs him. She is excited that he is alive and even more excited when I tell her that Sara is doing well. Kelly tries to start a conversation, asking about all that's happened, but I quickly quiet her.

"We made it to their group. As you can see by Chris being here with me," I say.

"Are they all–" Mary goes to ask something, but Julie interrupts before she can.

"Where is everyone else?" Julie asks. I can't help but notice a change in Julie's attitude. I know when my wife is pissed at something. Something had to have happened between them while I was gone.

"They have 78 people safe inside Paradise Cove. They are all scared to go out because Logan has gone on a killing spree. Evidently, he is killing anyone who questions his authority," I say.

"Logan has gone mad," says Chris.

"So, are they comi—"

"We're ready to leave if we need to," says Julie, cutting Mary off once again.

"No, they are coming here. We don't have much time, but I'm going to need all of you to make enough space for them. Clear out all the rooms so they have somewhere to go," I say.

"Okay, we can do that," Mary responds quickly. Julie rolls her eyes and looks away. *What is going on with them?*

"We met Karissa while we were out there," I say, pulling her in front of me.

"Hey, sweetheart, how ar–" Mary is cut off again.

"Who is she?" Julie asks. I grit my teeth so hard I can hear my molars squeak.

"Julie, can you stop cutting Mary off every time she speaks. What has gotten into you?" I snap. She doesn't answer. Instead, she huffs a deep breath and stares at me.

"She is going to help us with Logan. She and her brother

have been with Logan's crew and she agrees that he needs to be stopped."

"What happened to your face, Karissa?" Mary asks, staring at Karissa's swollen eye and split eyebrow that's dripping blood.

"I'm going to be a double agent and report back everything that is happening so we can be one step ahead of Logan and then take him out when we get the chance. To make it look convincing, I let Theo punch me in the face," answers Karissa before anyone questions why I hit her.

"How do we know we can trust you?" Mary asks.

"Yeah, how do we know you're not doing the same thing for Logan and telling him all of our plans?" Kelly asks, crossing her arms in front of her.

"You can trust me. I want my brother safe, and Logan is going to end up killing everyone if he isn't stopped," Karissa answers pointedly.

I hope we are making the right decision by trusting her.

"Our plan is for Karissa to return and lie on the ground beside Brent."

"Who is Brent? Julie asks.

"He's with Logans group and was patrolling the hallways with Karissa. I knocked him out, but we need to move before he wakes up." I look over to Karissa. "Karissa is going to go lie down next to him, so when he wakes up, he will think they were both attacked."

"Okay, how is she going to talk to us without anyone noticing?" Julie asked.

"She is going to leave a note in the closet where Tyler and I found the axes. She is going to explain that the closet needs to be checked nightly because that's the only place they hadn't checked, and that must have been where the attacker—me—had to have been hiding. The plan has some holes, but it's the best we could come up with."

Everyone looks at each other, but no one offers any other suggestions, so I carry on.

"Before all that starts, we need to get these people down here. Sara, the group's leader, is organizing things on their end," I say.

"Yeah, that makes much more sense. Sara is more of the take-charge type, always has been since we were kids," says Kelly.

"Chris and I'll head back down to Paradise Cove to get everyone. We'll be back as soon as we can. I'll knock three times so you know it's us. Be ready."

Chris, Karissa, and I quietly make our way back to Paradise Cove, constantly checking our surroundings as we go, even though we were told only Karissa and Brent were on lookout. I'd rather be safe than sorry. When we approach the hallway where Brent is still knocked out, Karissa quickly gets into position and lies near him. I consider the possibility that it's already been almost fifteen minutes, and Brent might wake up when we are relocating Sara and her group. I have to decide whether to continue or wait until Brent wakes up and leaves. *What do I do?* I realize that waiting could prompt Logan to attack sooner than he would seek revenge for us attacking his people. *We can't wait.* We need to move now.

Paradise Cove remains dark and as Chris and I approach, the guards recognize us, and taps on the door, signaling to remove the barricade. When the door opens, the musty smell and body heat wafts over me, nearly taking my breath away. I can't help but grimace at the stench. Inside, everyone is ready to move. I quickly instruct Sara to walk with me and tell Tyler to stay at the back of the line to be the last to exit. Together, I silently say a prayer that we are not ambushed by Logan's group.

We move as a group up the stairs in single file. Sara, Chris, and I make our way past Karissa and Brent. Karissa opens one eye and lifts a hand to us, and I nod my head to acknowledge her.

"What the hell happened here, and who is she?" whispers Sara.

"I'll explain once we're inside and safe," I say, glancing at Brent's body again. *Please don't wake up yet.*

"What about the girl? She opened her eyes," Sara asks.

"I'll explain once we're inside and safe. We're on the same team now, trust me."

"I don't trust anyone," says Sara.

This must strike a nerve with Chris, because he quickly responds, "Except for me. Listen to Theo; we can trust him."

Once we all arrive at the medical center, I knock three times. Shuffling begins behind the door, and the barricade is removed. Julie opens the door and lets us in. Off to the side, Mary stands while Kelly directs the people to sit along the walls until the doors are closed again.

Moments later as I'm helping Julie usher in the last of Sara's group, Tyler starts to follow then pauses an arm's length away.

"Tyler, what are you doing?" Julie says in a hushed tone. Tyler doesn't respond. He just stares down the hallway in the direction we just came from.

"Come on, bro," I say. Again, he doesn't respond. Julie reaches out, grabs his arm, and pulls him inside then quickly before she closes the door and secures the barricade again.

23

LOGAN

The deck doors of the Stardust Lounge are open and the cool, fall air is blowing gently inside, while the moonlight through the clouds casts the deck in dim light. People mill around, taking sips from their water bottles and chatting quietly, almost like they're on vacation rather than having been trapped on this godforsaken ship for months. It's a testament to my leadership of this group that they all feel safe. No one here is worried that we'll be overtaken by anyone from that bitch Chris's group, so while they all hide and cower like the little roaches they are, scurrying around on the ship, we do what we please. It feels damn good to be king.

Outside the main entrance, two guards stand with makeshift wooden spears. We take all the necessary precautions for safety here, having made several weapons, so the door to my lounge is open, unlike the doors at Paradise Cove. No one dares approach us without risking an attack from my army, which is as it should be. You fuck with the bull and you get the horns, and at this point, we have no issues taking anyone out who gets in our way.

I walk through the space and look around at people in various stages of conversation. The mood is pretty light, but I'm about to fuck that all the way up. These people have been slacking and they need to know that. Everyone needs to pull their weight here or they can get the fuck out. It's come to my attention that there's been some dissention in the ranks, and I'm about to nip that shit right in the bud. My word is the only one that carries any weight here, and it's been a while since they've had a reminder of that.

I make my way over to the center of the room where there's a large marble table that I use as a makeshift stage. I hike myself up onto it then look around the room, scratching my chin. My beard has grown in scruffy and thick in the time I've been on the ship, and I know I look gruff as hell and intimidating as fuck, but it works for me.

I stomp my foot on the table, pulling out the butcher knife I keep on me as a weapon. "Everyone, listen up!" I wave the knife toward the ceiling.

Most of the people turn toward me but the room stays noisier than I'd like. No, this won't do. I grab one of the empty liquor bottles on the table at my feet and slam it into the metal trash bin on the floor. The bottle clatters loudly in the bin, making a loud and obnoxious clang that reverberates in the room for a few seconds and gets everyone's attention. Instantly, the rest of the room falls silent.

"When I speak, you shut the hell up. Focus. On. Me. Am. I. Clear?" I punctuate the last of my words with a jab of my knife.

A quick round of "yes, sirs!" rings throughout the room. Some might just be going along with the others, but I don't give a shit if any of these people respect or even like me, I am the *only* person here that's keeping them safe and alive, so they will shut the fuck up and listen when I speak.

A man in the back of the room continues speaking, laughing at something. *The fuck?* I lean forward to get a better look, and when he fully comes into view, I see it's Jackson. He's a young and cocky little bastard who's been giving me shit for weeks now.

"Is something amusing to you? Did I say something funny?" I squint at him, daring him to speak. Jackson remains silent, staring back at me. I jump down from the table and stride toward him, and the crowd parts to give way. I stop right in front of Jackson, who's risen from where he was sitting, standing so close that I can smell his nasty-ass breath.

"One more time: did I say something funny?" I point my knife directly at his nose. If I leaned forward just an inch, with a flick of my wrist, I could cut it right off his smug-ass face.

"No, sir, I... I'm sorry. It won't happen again. I promise."

I stare him down and a gleam of something that looks like defiance flashes in his eyes. *Oh, hell no.* He lets out a breath that blows across my face and the smell of old coffee and rotten meat makes my stomach churn. I step back and take a breath, clearing my mind. I can't let this stand. If I let these people think they can get away with anything, to continue to defy me and not have clear loyalty and allegiance to me, they'll do everything they can to keep that shit up. Yes, I have all the power here, but my real strength is in keeping these people under my thumb. I have plans for this ship, I can't...no. *Focus, Logan.*

I take a deep breath, then exhale and count to five. Jackson's shoulders drop, and he shifts away from me, clearly thinking our conversation is done. Then I strike, thrusting my butcher knife into his stomach, twisting the handle deeply. The ten-inch blade pierces his gut and Jackson's eyes fly open wide. He looks at me, mouth open, as his gaze slowly drifts to my hand, still gripping the hilt of the blade.

"Why?" he chokes out, his voice garbled as blood spills from his mouth. The room is deadly silent, with nothing but the sounds of him coughing up the blood.

"You did this to yourself, Jackson." I twist the handle again and slide the knife from Jackson's torso, then swipe the blade against his shoulder, wiping the blood onto the fabric of his shirt. He drops to his knees, struggling to breathe. Gurgling continues to fill the silence, followed by a thud as his body slumps to the floor.

"Please...hel–" Jackson reaches out a hand and someone to his left makes like they want to help him and I look at them, glaring. Ah, Samuel. He's a good kid, but Jackson needs to be made an example of, so I can't let him help. Otherwise, these people will *never* learn.

I point my knife at Samuel. "No, Sam, I said *he* did this to himself. Leave him be. Anyone, and I mean *anyone,* who helps him will suffer the same fate. Is that what you want?" Samuel stills in his approach and just stares at Jackson, who's still begging silently for help. I ignore them both and make my way back to the table. I hoist myself back onto it, then survey the faces of everyone.

"Anyone else got something to say or think something's funny in here?" I look as many people in the eye as I can. I have no time for bullshit with these people. No one says anything, and I think I hear someone sniffling somewhere in the room. *Good.*

I think I still hear Jackson gurgling, and it pisses me off. That fucker's *still* trying to take the focus off me. "Drag his ass onto the deck, I can't hear myself think," I growl, pointing to the double doors. Two guys scramble forward and grab Jackson by the arms, then the crowd parts to allow them to drag him outside. A thick bloodstain smears across the floor behind them, and I can barely hold back a smile looking at it. I couldn't have planned this any better. The blood might stain the floors, but it'll be the perfect reminder for them: keep fucking with me and suffer the consequences.

As the guys return, I take a moment to gather my thoughts

before addressing the crowd again. They all seem to stay clear of the blood smear, and that's fine with me. Might as well keep it on display.

"Tomorrow night, we'll go to the Cove and get their water." I pause, my gaze sweeping around the crowd.

"But how? It's not like they're just gonna give it to us," Sam says, and a quiet murmur rumbles around the room. No one is flat-out disagreeing with me, because they *better fucking not*, but I don't like Sam giving them a reason to doubt my plan.

My jaw clenches, and I feel a bit of irritation rolling through me as I grip the handle of my knife tighter. I'm going to have to keep an eye on that one. Questioning my plans is damn near as bad as disagreeing with them. "I didn't say we were gonna knock on the door and ask for it like a cup of fucking sugar, Sam. We're taking it."

The majority of the crowd erupts in a roar of raucous sound, with some stomping their feet and others banging on the tables. Some sit silently, likely terrified at what they've seen, but I don't give a fuck about them. Just so long as they stay in line, I won't have any problems with them. I take a quick mental note of who they are to store that information for later, then focus on the energy of everyone else in the room. It's palpable and just what I've been craving. Many of us are like-minded and want order, craving complete obedience on this ship. And we will do everything we can to get it. *These are my people.* I feel a surge of confidence fill my chest and I stand a bit taller, grinning.

"Everyone, get your weapons together. Whatever you can find, grab it and get it ready," I say, and the boisterous crowd gets even more rowdy. I raise my knife in the air, joining in, and happen to catch a look at my hand, still covered in Jackson's blood. It's almost surreal, seeing his blood smeared on the back of my hand, almost like I'm watching all this through someone else's eyes. Like a dream. I feel a smile spread across my face, and my heart starts beating fast, the adrenaline flooding my system. It feels so damn good. I'm. Fucking. Invincible.

I look to the deck where Jackson's body lies motionless. I don't know if he's dead yet, but he's probably close to it, with all that blood pooling beneath him. It seems to flow slowly, like a red river, like it's making its way to the center of the room, like it's coming directly to me. I walk over to where Jackson is and crouch down next to him. He's breathing shallowly, and his eyes are closed. He won't be alive much longer. The blood pool underneath him is huge. Behind me, I hear everyone still moving around, still preparing themselves, and a buzzing in my ears drowns out the noise. I touch my fingertips to the pool of blood, then rub my fingertips together, feeling the sticky smoothness of the cooling liquid. I run my fingers up the side of Jackson's shirt, then dip them directly into the stab wound. I dig around a little, and Jackson lets out a weak groan, shuddering a bit.

I lean forward, putting my lips almost to his ear. "You wanted my place here, didn't you, Jackie-boy?" I dig my index and middle fingers into the wound a bit deeper, and he groans again, barely moving this time. When I pull my hand back, I smear some of the warm blood down the side of my cheek, then position myself in front of his face, where his eyes are now wide open from the pain. He blinks slowly as he stares at me.

"You could never be me, and you sure as shit can't beat me," I say as I watch the light in his eyes start to go dim. "And now, you'll forever be part of the reason why *everyone* here will stay fucking terrified of me. So, thank you for that."

Jackson gazes at me silently, and I see one last bubble of bloody spit burst on his lower lip before he lets out his final breath. I dip my fingers in the pool of blood and mark a bloody cross on his forehead, then I drag him over to the side railing of the deck, and heft him up and over, tossing his body into the dark abyss. I stand to my full height and take a deep breath of the cool night air.

"Hello... sir?" I spin around, my hand instinctively flying to my knife as I'm startled by a voice behind me. "Is...is everything okay?" A young man– Alex, Andrew, I think? – blinks when he sees the blood on my face. Everyone else is still bustling around the room behind him. Most of them are now breaking chair and table legs and such to make weapons.

"Yeah, I'm good. I was just tossing out the trash," I say, then scratch the back of my neck. "What's on your mind, son?"

"My sister Karissa hasn't returned from her shift yet, and normally she and Brent are back by now. I'm worried something might have happened." He shuffles on his feet, dragging a hand through his hair. Huh, I hadn't even noticed they'd hadn't returned.

"I wasn't aware Karissa had a brother. What's your name, son?"

"Xander, sir," he says, frowning. "I'm worried because they usually don't take this long."

I think for a moment. Huh. Come to think of it, the kid's right. I stride past him into the room and let out a wolf whistle. All heads turn toward me, and a few people gasp. I assume it's due to my bloody face, but I don't give a fuck what they think right now.

"Listen up, people," I say, taking my position back on the marble table. "We're missing Karissa and Brent. They haven't returned from their patrol, so change of plans." I look pointedly around the room. "We're moving out tonight. Prepare yourselves." Everyone scurries back around, and Xander walks up to me, eyes wide in surprise.

I put a hand on his shoulder, and he flinches when he sees the smear of blood I leave there. "We'll find them, then ambush the Cove and take whatever else we want. Anyone who stands in our way dies tonight. If you can't handle that, you need to make a decision now, son." I look him deeply in the eyes, letting him know in no uncertain terms that there is only *one decision* to make.

"Yes, sir," he says, taking a breath, then walking back to the far side of the room. He picks up a paring knife and a screwdriver and slips them into his pocket.

"Listen up, everyone!" I shout to get everyone's attention. "Walk outside and smear a stripe of Jackson's blood on your face. I want to know who my people are when we attack." For a moment, I think they are not going to listen. I see a few faces cringe at the order, but they don't disobey; they do as they are told. I can't help but smile. It's so fucking good to be king.

24

BRENT

The hallway is dark when I open my eyes. My head is throbbing. I reach up and gently rub my scalp, wincing at the pain. I feel something sticky and wet, along with a sting of pain. I try to see if there is blood on my hand, but my vision is blurry. Squinting and blinking repeatedly, my vision begins to clear, and my hand slowly comes into focus. Moonlight streams through the window, revealing that the wet substance on my hand is thick and dark red. I reach up again and gently run my finger over the wound, trying to gauge its size, and wince again from the pain. *What happened? Where am I?*

As I look around, my eyes land on a woman lying on the floor across from me. I squint and blink a few more times, my vision going in and out. I stand up and hold myself up against the wall. A wave of dizziness takes over, and I nearly fall. After a moment, I drop back to the floor and crawl closer to get a better look at the woman. My mind feels fuzzy, and I'm unsure whether what I'm seeing is real. I close my eyes to steady myself. Several seconds pass, and memories of earlier begin to return. *Karissa.* I remember who she is. My patrol partner.

"Karissa, wake up," I whisper, shaking her shoulder, but she doesn't wake up.

Looking around the empty hallway, I don't see anyone. I can hear people whispering and footsteps in the distance, but I don't know where or who they're coming from. I go silent, desperately trying to pinpoint where the sound is coming from. My ears start to ring, and a sharp pain throbs in my head again. I lose the sound of the footsteps. Quickly, I stand again, holding the wall and stumbling for a moment. I then rush to the stairwell, following the sound and trying to listen again. This time, the whispers and footsteps are clearer— "Go quickly before they see us, get inside," says a man's voice I don't recognize.

Quickly, I climb the stairs, not caring if my vision is fading in and out or if my ears are ringing nonstop. With every step, I'm closer to the voices and footsteps. I need to know who did this to us. As I step onto the 11th floor, I turn into the hallway facing the medical center. There's a figure, but I can't tell who it is without them turning around.

"Hey," I shout. The figure stops and faces me. His face is familiar, but I can't quite place who he is; my vision is blurry. I blink a few times, and my vision clears. It's Theo. No. That's not right, something is telling me it's not him. Maybe the way he is standing. Then it hits me—it's Theo's twin brother, Tyler. I stand silently, staring at him. He sees me too, because he stops moving, locking eyes with mine in a tense stare. Neither of us moves a muscle. Tyler's gaze falters when someone steps out and grabs his arm, pulling him inside the medical center.

I take off running to catch him, but by the time I make it to the medical center's doors, they are closed. I almost fall over. I'm so dizzy, and my vision keeps going in and out. I reach toward the door, but I'm too late. I can hear things sliding against the floor and being slammed against it. Still dizzy and out of breath, I fall backward, landing on my back and nearly knocking the breath out of me. I stare at the ceiling, taking slow breaths until I'm able to sit upright again. My mind is racing. Karissa's body flashes in my memory, and I slowly get up, trying not to make myself dizzy again. I rush toward the stairwell and race back down. Karissa is gone—nowhere to be found.

I look around and realize the blood staining the carpet where I was lying a moment ago is also gone. I glance up and see that I'm on the wrong floor. Frantically, I race down one more flight of stairs. Then I see Karissa—she's not moving. My first thought is that she's dead. Without thinking, I run and drop down beside her, beginning CPR. The instant I press my weight down on her torso, she screams in pain, and her eyes fly open, fear evident. She sucks in a deep breath and shoves me off.

"What are you doing? Get off me. Are you trying to kill me?" Karissa shouts.

"I thought you were dead." I answer, confused about everything right now.

"Well, I am alive, but now I feel like I am going to puke," she says, grabbing her head and then lying back down on the floor. "Ugh, my head is killing me. What happened? Where are we?"

"I'm not sure, but I think we were ambushed. I saw Tyler go into the medical center upstairs. I tried to catch him, but I wasn't quick enough."

"Are you sure? They haven't opened that door in weeks."

"I know what I saw... or at least I think so." I rub the back of my neck.

"Let's go. We need to tell Logan."

"Can you remember anything?" I ask, my memory hazy.

"I don't know. I remember walking and talking with you. That's all I remember."

"We checked everywhere. Where could they have come from?" I ask, scanning the area. Karissa remains silent as I follow her gaze and notice she looks at a closet we didn't check, then past it. A second later, she takes off running toward the stairwell and stops, turning to face me.

"What are you doing?" I ask. She doesn't answer. Instead, she begins walking, and I realize she's retracing our steps. Then she stops halfway.

"This is as far as I remember going. I remember hearing something and asking you to check it out. The closet is the only place we didn't check," she says.

"Let's go tell Logan before anything else happens."

We start walking down the hall toward Stardust Lounge, which is at the opposite end of the ship from where we are. Before we reach the entrance, a mob of people with their faces and arms smeared with dark red paint appears. It's our people, it's Logan's group. Why do they have face paint? Then I look down at my hands and realize the color is exactly the same. I grab Karissa to get her attention.

"What the hell is going on? Where are they going?" I ask.

"No idea. Let's get out of their way before we're trampled."

The crowd rushes past us, screaming and chanting. I watch carefully as they run past me.

"They must be going to Paradise Cove, but the ambush wasn't supposed to happen until tomorrow night," I say.

"Were we knocked out that long? Is today... tomorrow already?" Karissa asks. I stare at her, confused. How long had we been knocked out?

"Something must have happened," I say.

The crowd continues rushing past. Once the commotion has calmed, Logan and Xander emerge and walk by. Logan's face has a single red stripe running down it. Xander's face also has a crimson stripe. Karissa looks up and I see her eyes go wide, then she takes off running toward them. I've overheard her and Xander talking in the past and put together that they're related. Xander sees Karissa rushing toward him, and they collide in a hug. Logan stares at both of them with a puzzled expression.

"Looks like your sister is just fine," Logan says, his tone cold.

"Yeah, it looks that way," replies Xander. Karissa steps back and meets Logan's eyes.

"I'm not okay. Brent and I were ambushed. How long have we been gone?" she asks. Before Logan has a chance to speak, I walk up and take his attention.

"She is right. We were ambushed and knocked out. My head is killing me. I have a gash where someone hit me."

"Who did this?" demands Logan.

"Tyler, at least I think it was him," I answer. Logan looks at me like I have horns growing out of the sides of my head.

"Tyler, as in Theo's shit-for-brains brother? You mean to tell me that little asshole knocked both of you out and got away?"

"I guess. I don't know," I answer, not knowing how else to respond. I look over at Karissa and she doesn't say a word.

"I call bullshit. How do you know it was him?" Logan asks.

"Because when I woke up, I heard people talking, and when I followed the voices, he was there, at the door to the medical center. There was something about the way he looked at me." I don't know it for sure, but he must've had something to do with it.

"So you mean to tell me that Tyler snuck out, attacked you, then ran away? That doesn't make any sense," says Logan, and I have to agree; he has a good point. It does sound strange.

"Well, when you put it that way, it sounds dumb," I answer.

"Damn straight. It sounds fucking ridiculous! Why would he do that?"

"I'm not sure, but it's not just the five of them inside there anymore, I think—"

A woman's voice shouting cuts me off. I quickly turn to see her screaming, running from the direction of Paradise Cove.

"Logan! Logan!"

"What?" answers Logan.

"Paradise Cove is empty. They're all gone," says the woman.

Logan slowly turns back to me, and at the same time, we both say the exact same thing.

"The medical center."

25

THEO

The medical center is crowded as people search for a place to sit. It's clear that everyone is exhausted and out of sorts as the night goes on. I make my way to the back closet, where Mary is organizing the items on the shelves. I ask if she needs any help and she waves me along, telling me she can handle it. As I'm walking away, Sara approaches to speak with Mary. While Sara and Mary talk, I make my way over to Julie, who is off to the side chatting with people. I look around to make sure no one is listening, and I notice Kelly and Tyler are only a few feet away, moving people around to help them get settled in. Not wanting Kelly or Tyler to hear me, I pull Julie in the back, where no one else is around.

"What's going on with you and Mary?" I ask as soon as we are alone. She gives me a confused look, as if I just asked her to jump off the side of the ship.

"What are you talking about? Nothing is going on between us. Why do you ask?"

"It's obvious something is going on between you two. We've been married long enough for me to know when something is bothering you. Like earlier, when you wouldn't let her speak, you kept cutting her off. What was that all about?"

"It's nothing. Really, we are fine."

"Then explain why you wouldn't let her speak."

"I did let her speak," she answers, her brows furrowed as she stares at me. The tone in her voice tells me I've hit a nerve, which only confirms my suspicion.

"No, you didn't," I say once again.

"Okay, maybe you're right, but ever since we've all been stuck here, she acts like she runs the damn place, and it's getting on my nerves."

"Julie... she *does* run the place. This is the medical center, and she is the only nurse on board," I say.

"Okay, but that doesn't give her the right to speak to me the way she does. It's my last name on this ship, not hers." I don't respond. I stand silently, stunned by her words. This is a side of Julie I've never seen before. In the past, I never really paid it much attention when she'd get like this; it would just blow over after a while, and she'd just move on to something else. Like when my dad left the business to Tyler and me. She swore up and down that he was too irresponsible and she wasn't going to let it happen, but it really wasn't up to her. Dad could do whatever he wanted with the company, whether she liked it or not. In the end, it didn't even matter, since he'd already signed over his company shares to me. Still, her words *It's my name on the side of the ship, not hers* rings in my ears and pisses me off. Why does it even matter at this point?

"Actually, my name is on the ship, Julie. I'm the one in charge, not you!" I snap. *Oh fuck, I've turned into my father. That's something he would say, not me.*

Shock washes across Julie's face. Before I can apologize, she turns and walks away back to Tyler and Kelly. I think about following her, but I make my way to the center of the room instead. The sun is beginning to rise as the early-morning's faint, warm glow breaks through the clouds, and I can see the toll that being trapped not only on the ship but also in a confined room with little water and food has taken on everyone. I don't want to keep going like this; from the dirty clothing to the unshaved faces and unkempt hair, I realize that if we don't change what we are doing, no one will survive, and Logan will end up killing us all.

"Can I have everyone's attention, please?" I say.

Everyone in the room turns to look at me. I watch Mary and Sara emerge from the closet. I don't want to cause more drama, so I call Sara up to stand with me in solidarity. She joins me in the center of the room, waiting for me to speak.

"We can't continue living in fear of being killed or not surviving. Winter is quickly approaching, and we don't have enough food or water to stay separated. We need to work together to survive. I know it's been months, and I can imagine many of you have given up hope that we will be rescued." I pause and look around the room at all the faces looking back at me. Sorrow, defeat, and hunger is all I see. I know deep down I need to give these people an ounce of hope, or else they'll give up entirely. I take a deep breath and continue. "I need you all to believe that we will be rescued. I have no idea why no one has found us yet, but I am sure that when we get home, your families are there, waiting and clinging to the hope that you are alive."

"Theo is right, guys. We are going home. Maybe not today, but tomorrow or the next day. Theo, what do we do?" Tyler pipes up from the back of the room.

"We have each other's backs and protect each other when Logan does show up."

"I always have your back," he says, putting a hand on my shoulder. His words remind me of when we were kids, when I was the one protecting him. There wasn't anything I wouldn't do for him—then and now. And when the roles are reversed, I trust him to do the same.

"Thanks, Tyler. I know everyone is exhausted, but Logan is going to realize soon that no one is in Paradise Cove anymore and he and his group will come looking for —" My words are cut off by banging on the door. A collective gasp sounds in the room.

"We know you are in there. Open this fucking door!" Logan shouts.

Standing tall, I turn my attention back to the crowd as Logan keeps banging on the door. Everyone's wide-eyed and completely alert now. "Everyone, grab your weapons and do your best to stay alive. Leave Logan to me." Heads turn, looking around at each other.

Then Sara speaks up, breaking the silence, "Don't just sit around. Get your asses in gear."

At her command, the room erupts into chaos, and everyone starts running around frantically, grabbing the makeshift weapons they've created. There is a constant banging on the door as Logan's group attempts to breach it. Tyler shouts to get Kelly's attention, but she doesn't hear him. I see him searching, and when he finds her, a smile spreads across his face as he watches her pull out a wooden broom handle with a broken glass bottle attached to the end. The thick, sharp glass fragment was held in place by layers of duct tape. On the opposite end, another piece of the broken handle is taped together like a cross, with butcher knives taped to the three ends. When she looks up and sees Tyler staring, she smiles back and winks. He quickly runs over to her. They're nearby now, and I can hear their conversation.

"When, how, and where did you get that?" he asks.

"Julie found a case of butcher knives in the kitchen a while ago. She gave them to Mary and me and kept one for herself. I found the duct tape and broom in the supply closet the other day. I haven't been able to sleep, so at night I've been going into the supply closet to make this," Kelly answers.

"That's badass, babe."

"I made something for you, too."

Kelly runs back into the closet and brings out a hammer with a butcher knife duct-taped to the hammer's handle, with a makeshift sleeve made of silver duct tape. Tyler grins as big as his face as soon as Kelly hands him the weapon.

"Holy shit, babe! Thank you," Tyler says.

"Now give that axe to someone else who needs it." Tyler goes to give the axe to one of the people nearby, but Kelly stops him. "There is one more thing I think I need to tell you," she says, now with a more nervous look. I can't help but inch a little closer to eavesdrop on what she is going to say.

"Can it wait until later? Logan is about to break the door down," Tyler asks.

"No. I need to tell you now in case we don't make it," she answers.

"Don't say that," Tyler says, pulling her close. "We are going to make it."

"I'm pregnant," Kelly says.

Tyler stands there, silent. The banging on the door gets louder.

"Babe, did you hear me? I think—"

"I hear you. I'm trying to process what you just said," Tyler answers, cutting her off. "It's okay. We will figure it out. I love you." Then he kisses her.

I want to run over and give my brother a hug and tell him congratulations, but then I think about what it means if Kelly is pregnant, and my heart starts racing—this is not good news. This is a death sentence for that baby. There is no way a baby is going to survive on this ship. We are barely surviving. I feel a tear forming in my eye at the thought. I wipe my eyes and see Tyler staring at me. He doesn't move, just stares. The sound of banging pulls me from my thoughts, and I turn away from Tyler's gaze.

"Theo, open this fucking door, or I will break it down and kill you myself." I run up to the medical center's entrance. Tyler, Julie, and Kelly stand beside me, and an entire army is behind us, ready to fight. I smile at Tyler, holding up my axe, and he turns to Kelly.

"Shit is about to get crazy," he says. Kelly smiles, gripping her weapon.

"I'm ready," she says.

"Undo the barricade," I shout.

The door opens, and Logan and his group surround the entrance, staring directly back at us. Everyone falls silent as Logan and I stare at each other, waiting for someone to make the first move. I don't want to wait; I break the silence.

"I'm going to kill you, Logan."

All at once, both groups start shouting and running past Logan and me, their weapons raised, and the battle begins.

26

THEO

Within seconds, Logan disappears into the crowd and I lose sight of him. Desperately, I look around the room to find him. Then I'm shoved from behind, and I feel a sharp pain as a knife slices the side of my arm. Blood pours out. Quickly, I realize that not only is Logan trying to attack me, but so is the rest of his group.

I scan the room and notice that everyone in Logan's group has blood streaked across their faces. I shift my attention and see Tyler running past me. I reach out and grab him.

"All of Logan's people have blood on their faces. Tell all the others so we don't attack our own people." Tyler looks around the room, then he gives me a thumbs-up.

Mary stands off to the back of the room near the closet, behind a table leaned up against the closet door, holding a knife. I also see that one of Logan's people has spotted her and is about to attack, but she jumps from one side of the table to the other and stabs at her attacker. I quickly run over and make eye contact with Mary. Her knife isn't doing much to protect her against the long spear coming at her. It's obvious that she is becoming exhausted from the constant jumping back and forth.

The bloody face doesn't realize I'm behind him. I lift my axe above my head and swing. The axe lands directly in the attacker's skull and splits it in half. Blood splatters across Mary's face. The body falls to the ground and I quickly pull off my shirt and toss it to Mary.

"Clean the blood off your face," I demand. "Logan's group is all marked with blood across their faces. Anyone with red across their face is a target."

Off to the side of the room, a group is battling. As I glance over, I catch sight of one of my people being stabbed in the eye with a broken bottle, then stumbling backward. I hear a scream from somewhere else—I recognize it immediately as Julie's. I quickly turn, searching for her. When I find her, I watch as my wife takes a butcher knife and jabs it into her attacker's face repeatedly. My eyes go wide watching her. Then, I watch as another bloody face runs up to her.

"Julie, watch out!" I scream. She looks up just in time to see the man rushing at her with a broken bottle. She quickly jumps out of the way as he rushes past. Julie swings in a quick motion, slicing the man's Achilles tendon. Instantly, he drops to the ground, grabbing at his ankle. Julie leaps on top of him and drives the knife directly into his throat. Blood sprays out, and his body goes limp.

A woman runs up to me with a small knife in her hand. I step aside and swing my axe just as she gets close enough. My axe lands in her throat. She grabs at her neck as blood gushes out. I watch as she opens her mouth to speak. No words come out, only gurgles. I turn to see another person charging at me. I throw my axe and it lands in the man's chest. He stumbles backward to the ground. I run over and grab my axe. Then, as he is lying on the floor screaming, he gets trampled by countless people. Tyler and Kelly come into view, and I watch them fighting their own battle with a group of bloody faces.

Tyler slams his hammer down on a bloody face's collarbone. There is a loud crack, and the bloody face drops to the ground, screaming in pain. Before his attacker can react, Tyler spins his hammer and uses the blade to slit the man's throat. I can't help but smile at my brother being such a badass, protecting and fighting off everyone trying to kill us.

I turn to fight off more of Logan's group and stop when I hear Kelly scream. I turn back to see her against the wall. A man backs away from having her pinned against the wall. I gasp when I see the man pull a knife out, and crimson blooms under her shirt. Tyler screams, his face red and full of fury. He ambushes the man from behind, pulling him to the ground. Quickly, the man twists his body, putting his back flat against the cold hardwood floor. Tyler leaps on top of him like a maniac, screaming so loud that saliva spews from his mouth.

Full of rage, he lifts the steel hammer and slams the blunt end repeatedly into the man's face. One swing after the other until his skull is smashed in, leaving nothing but a bloody cavity. Kelly's eyes are locked on Tyler as he exhausts himself swinging. She lifts her shirt in a panic, revealing where the bleeding is coming from. I watch as relief washes over her when she notices the knife missed her stomach and pierced her right side. She puts pressure on her wound with her palm and t-shirt, then runs to Tyler, pulling him off the man he just killed. I hear her screaming at Tyler that she and the baby are okay.

Tyler looks up to her then falls to the ground from exhaustion, but as he lies on the ground, another bloody face tries to ambush him and Kelly. Tyler must also see him coming because he grabs his hammer with the knife end out, and slams it directly into the guy's chest. He drops to the ground, and Tyler drops his arm, letting out a ragged breath. I'm ambushed again, turning my attention back to fighting. I hear Tyler scream and I turn to see another bloody face kicking him in the ribs. I turn around and swing my axe at the woman attacking me. My axe slams into her chest, causing her to drop to the ground. I rush over, tackling the man kicking my brother, then I crawl back to Tyler and Kelly.

"I'm fine. Check on Kelly, please," shouts Tyler.

"I'm okay," answers Kelly.

While I'm on the floor checking on Tyler and catching my breath, suddenly I feel excruciating pain as a blade rips through my right shoulder and exits through the front. The pain is white-hot. As if in slow motion, I turn my head and see my reflection in the blood covering a large butcher knife. Slowly, the butcher knife begins to disappear as it's pulled out. I turn my body to see who stabbed me in the back, but I already know exactly who the culprit is. I need to make eye contact with him before my life is taken.

When I'm facing Logan, he has saliva and blood dripping from his mouth like a rabid animal about to attack, with an evil grin spread across his face. He lifts his boot and shoves me backward onto floor. The pain in my shoulder is so intense that I can't lift myself. With both hands on his knife, Logan slowly lifts the blade high above his head, ready to slam it into my chest. I lie there, unmoving. But before he slams the knife down into me again, he leans down to whisper.

"I want to watch the life drain out of your eyes while you stare at me." He then spits a mouthful of blood in my face. "Your father would be so disappointed in the weak little shit you've turned out to be." Those words cut deeper than any knife ever could. Before I can react, Logan lifts higher to slam his knife into me. I close my eyes, bracing for the impact. I hear Tyler scream and then a gasp of breath from above me. The entire room goes silent. When I open my eyes, I see a broken wooden handle pierced directly through Logan's heart. On the opposite end of the wooden handle above me is Tyler, dripping with sweat and out of breath.

"My turn to protect you, brother," he says, looking at me, then collapses across my chest.

The chaos and killing in the room stops as everyone watches Logan take his last breath, then fall backward, his face showing shock at his defeat. I remain still, having just watched my life flash before my eyes. Taking a deep breath, I muster all my strength and roll Tyler off me to check his condition. As I check on him, he begins to blink and regain consciousness, and I bend down, getting close enough that only he can hear me.

"Thank you," I whisper, grateful.

Kelly rushes to Tyler's side, while Julie stands silent with a shocked expression.

I give Kelly a smile, then lean in closer to her. "Congratulations on the baby."

She gives me a surprised look and opens her mouth to say something. I cut her off and answer before she can say anything, "I heard you tell Tyler."

Kelly smiles back at me, then bends down to check on Tyler. Julie makes her way to me, still looking shocked by the situation. Mary throws the table she is standing behind to the ground, hitting the floor with a loud thump. The noise startles everyone, and we all turn to face her.

The bright sun shines into the medical center, illuminating the chaotic scene. Battered and beaten bodies are scattered around. Crimson drips down the walls and glass doors leading to the balcony, flowing like a slasher film. We all stare at each other in shock, unable to move or make a sound. I walk to the center of the room and stand calmly with Julie by my side. Tyler remains on the ground, with Kelly kneeling beside him.

Sara joins me in the center of the room and starts to speak to everyone. "Look at what we've done and what we've become.

All this fighting, and for what? From now on, we are one group. Together, we will survive until we are rescued from this nightmare. If we are resourceful, and if we work together and combine our ideas, we can find a way to survive without killing each other. If you don't agree, get the hell out now."

Before anyone has a chance to react, Karissa emerges from across the room screaming. Everyone turns to see her drop to the floor next to a young man. The young man is either dead or unconscious because he isn't moving.

"Karissa, what's wrong?" I ask.

Before she can answer, I look at the body she is holding; I don't need her to respond to know what she's going to say. It's her brother.

"Mary, we need help!" I shout.

Everyone steps out of the way as Mary rushes over and drops to the ground beside Karissa and her brother. Karissa looks up at Mary with tear-filled eyes. "Please help Xander. I can't lose him; he is my only family," Karissa croaks through sobs.

Mary looks down at Xander, examining his wounds. His stomach rises and falls, showing he's breathing. She doesn't seem worried about the superficial wounds all over his body; all her focus is on the wooden spear protruding from his left leg.

"If we are going to be able to save his leg, I need to remove the wooden stick and pray that it's not in a major artery. It's a good thing he's knocked out because this is going to be painful." Karissa and I both stare blankly at her.

"What can we do?" asks Karissa.

"I'm going to need help holding him down and I need a medical kit. There is one in the closet. Theo, can you grab it?" asked Mary.

"Yes, I'll be right back," I say.

I get to the closet, open the door, and notice a medical bag sitting on the floor. I quickly open it and see tons of medical supplies. I grab the bag, wincing in pain from my shoulder. I can't lift the bag because the pain is too intense, so I drag it over to her.

All at once, several people approach Xander and quickly kneel, offering their help. They each place their hands on different parts of his body, making sure he is secure.

"Okay, give me the iodine, suture kit, and some bandages," says Mary. I slide the bag over to Mary, not entirely sure I'd grab what she was asking for. She digs in the bag, then spreads out all the items next to Xander, and looks up at us all. "On the count of three, hold him with everything you've got. One. Two. Three." Then she pulls the wooden spear out. A portion of the spear has broken off under his skin, visible to everyone. Xander wakes up screaming in pain. His eyes are wide. His body flings around as he screams louder than I've ever heard a man scream.

"I have to remove the rest of the wood and suture him." She slips her fingers into Xander's wound, digging for the wood fragments. Xander screams again, flailing his body. Suddenly he stops, and his body goes limp, passing out again.

"He is going to be fine. He's just passed out from the pain," Mary says directly to Karissa, who looks like she is about to have a full-blown panic attack while watching her brother. Once more, Mary tells everyone to hold Xander down. She grabs an antiseptic wipe, cleans the wound, then grabs the sutures. Karissa turns her face away the instant Mary punctures Xander's skin with the needle. Within seconds, Mary has the wound closed.

"Quick! Someone, hand me the bandages before he wakes up again," Mary says, reaching blindly for them.

The roll of bandages is handed to her, and she quickly begins tightly wrapping the leg.

The room falls silent as she finishes, then backs away, wiping sweat from her forehead.

One by one, people line up with injuries, begging Mary for help. Of all the injuries I can see in the room, Xander's leg and my shoulder look to be the most serious; all the other injuries seem to be non-life-threatening.

I look over and see Julie standing alone in the back of the room, her gaze fixed on Mary. Kelly and Tyler stand off on the opposite side of the room from Julie, holding each other close. I turn to see Sara staring at Tyler and Kelly as well. I bring my attention back to Karissa sitting on the ground beside Xander, who has yet to wake up. Dead bodies are scattered all over, their crimson blood pooling on the floor. The remaining survivors stand around, holding their wounds and silently looking around. I stand up and make my way around the body to the center of the room. Logan's body is still warm at my feet. My body is sore, and I look down at my shirtless torso and see blood all over, both dried and fresh. The gash is still seeping blood from my shoulder blade. I'm suddenly dizzy and close my eyes to steady myself. After a second, the dizziness subsides, and I open my eyes to everyone staring at me.

"Get treatment if you need it, throw the dead bodies overboard so they don't rot, and let's do our best to stay alive." As I finish my sentence, my knees buckle, and I slump to the floor. I can't move. I hear Tyler shouting in the distance and feel someone shaking me violently. I didn't even realize that I closed my eyes. When I open them, I see Mary running toward me, dropping to the floor next to my brother. I'm too weak to move, so I stare off and focus all my attention on the first thing I see. It's Kelly. She's alone. I watch as she looks down and lifts the shirt she had placed on her right side. A moment later, she collapses onto the floor. I try to shout at Kelly, but nothing comes out. I'm too weak, and then, everything goes black.

27

MARY

"Mary, help!" screams Tyler while I'm in the middle of wrapping a bandage around an arm. I look up to see Theo sprawled out on the floor next to Logan's body. Theo is covered in blood. So much is happening at one time. From each corner of the room, I can hear my name being called. Someone needs me everywhere I turn. *I've got this. I've handled far worse than this. I work best under pressure.* These are the things I have to remind myself when I feel myself getting overwhelmed. I finish up with the bandage, take a deep breath and run over to Theo.

I drop down beside him and feel for a pulse. He is still breathing, but he has lost a lot of blood. I look up to find Tyler staring at me. "Grab the medical kit I used on Xander, and bring it to me." I point to the bag behind me. When he returns, I dump out all the supplies and grab what I need. A moment later, Sara screams in the distance, and I turn to see Kelly on the floor. I take a breath, then refocus on Theo. I can't think about anything else right now. Tyler, sitting here with me next to Theo, looks up at Sara's scream as well. He looks at me, then back to his brother. I watch as he pauses, and I can tell he is questioning what he should do.

"I've got this, Tyler. Go help her," I say.

He jumps up and races to Kelly's side, a few feet away. When he reaches her, he drops to his knees and lifts her head.

"Are you okay?" I hear him ask.

"Who are you?" Kelly shouts, her voice higher than normal.

I turn my attention to the commotion and see Kelly crawling backward from Tyler. Kelly glances around the room, visibly terrified by the people around her covered in blood, and then her eyes go wide when she sees all the bodies lying all over the room. She starts breathing rapidly and I can tell she is having a panic attack. Sara rushes over to help Kelly, but before she reaches her, Tyler tries to ease up next to her. Kelly reaches up and slaps him across the face.

"Stay away from me," shouts Kelly. "Someone, help me. This man is trying to attack me."

Tyler suddenly stops and starts backing away. He looks around the room and everyone else looks just as confused as he is.

"I'll handle this. Go help your brother. She'll be okay, I promise," Sara says, pointing to me. Tyler stares blankly at her, lost in his own thoughts.

"Come on, Tyler, she is right. Come help me," I say, motioning him over to me.

Sara runs to Kelly's side and whispers something in a hushed tone, then gently rubs Kelly's arm. My heart is racing with adrenaline. I look down at Theo's limp body. I only have so much time before he is dead.

"Tyler, I need you right now!"

Shock remains evident on his face as the events unfold and he stands motionless beside me.

"Tyler!" I shout again, loud enough to get his attention. He turns to face Sara, who is holding Kelly and calming her down, then looks back at me, glaring. He shifts his gaze elsewhere. I follow his gaze and see he is looking at Julie, who remains in the same spot as moments ago. She is staring back in our direction, her face emotionless. I can't handle Julie right now, I have to focus on Theo.

"Tyler!" I shout again.

This time he listens and looks at me. "What can I do?" he asks.

"Take this and clean your arm. We need to do a blood transfusion immediately. If Theo doesn't receive blood now, he is going to die. You should share the same blood type, making you his only chance for survival," I say, handing Tyler an alcohol wipe.

"Oh. Okay," he says, taking the bottle of rubbing alcohol and pouring it over the crease of his arm. I look at Julie and her face instantly creases into a scowl when our eyes meet.

Julie stands there staring at me; then she looks around the room at everyone watching her. Off to my side, I hear a group of people whispering and make out one of them saying, "Why isn't she moving?"

"Julie, what are you doing? Come over here," Tyler shouts.

Julie seems to snap out of her thoughts and runs toward us, dropping to the ground beside Theo.

"I'm here. What can I do?" She asks, then looks at me as if waiting for an order.

Think. Think. Think.

"He has lost too much blood. I can hardly feel a pulse. We have to start the transfusion now," I say. "Julie, in the bag is a bottle labeled Citrate Phosphate Dextrose. Grab it."

After frantically searching for it, she hands it over and I grab a needle.

"Quick, hold this bag while I place an IV in Tyler's arm," I tell Julie, handing her an empty IV bag. "We need to fill this bag with Tyler's blood and transfer it to Theo. But without this, it won't work." I say, holding up the medication and needle.

"What is that?" Julie and Tyler both ask.

"Anticoagulant, I have to mix this in the bag while it's filling with Tyler's blood so that it doesn't clot and is able to go into Theo. This is risky as hell, but by the grace of God and a miracle, it might work." Julie's eyes lock onto mine with a blank expression. I turn away and face Tyler. "Are you ready?"

"Yes."

"Julie, I need you to hold the bag while I inject the anticoagulant," I say, then insert the IV into Tyler's arm.

Tyler turns his head and clenches his fist as the needle pierces his skin. Blood begins to flow through the clear tubing, filling the bag. My shoulders drop in relief after being so tense. Tyler gives as much blood as he can without passing out like Theo.

"Is there anything I can do?" asks Karissa.

"No, I think we have it under control right now. Stay with Xander."

While the bag fills, I mix the anticoagulant. Then, I suture Theo's shoulder and insert an IV into his arm. Moments later, we all sit back and exhale, exhausted and drained.

"What do we do now?" Tyler asks.

"We wait. That's all we can do. I've done what I can. I'll keep an eye on Theo," I say, then look around at everyone else. "Thank you for your help."

Karissa smiles and responds, "Anytime. Thank you for helping my brother."

"This is going to take a while. Tyler, you may have just saved your brother's life." He smiles back at me, then looks over to Sara and Kelly, and his smile drops. I feel bad for the kid. Kelly may have lost her mind. At first, I thought she was having a panic attack but this seems much worse than that. I reach over and grab Tyler's hand. "She is going to be okay," I say, hoping it's true.

"Thank you," he whispers. His eyes are wet as he hangs his head low. I look around the room at all the people sitting around Theo. My eyes land on Julie, who seems lost in thought. She doesn't realize I'm staring at her, but I know exactly what she is thinking. I look back down at Theo. *This could have ended entirely differently. I pray this works, but only time will tell.*

28

TYLER

My mind is racing, and I can't think straight. I look down at my arm and see the IV, then at Theo on the floor. *Please wake up. I can't do this alone.* I can hear my father's voice echoing in my head, reminding me what a fuck-up I am. *I really hope he's wrong today.* I wonder what he would say if he knew he would be a grandpa? I'm sure he would find some way to put me down and tell me I'm not good enough. *I'll be a better dad than you, you old bastard.*

I turn to check on Kelly and a wave of confusion surges through me. She was so scared when I walked up to her, and then she shouted at me for attacking her. *Why would she think I was attacking her?* I push the thought aside, closing my eyes, and I take a deep breath. All the passengers are beginning to clear the room, with some still arguing about Logan being out of control and how they only did everything he said so that he wouldn't kill them. A moment later, my eyes shoot open when a fight breaks out in the back of the room. Two guys are going on about how Logan didn't deserve to be killed. I look over and see Brent, the same guy that I saw in the hallway earlier.

"Logan was right. You're all going to die out here!" he spews.

A few others stand and shove him out of the room. Karissa is yelling at him as he is being pushed out, but I'm not sure what she says. For now, all I want to do is close my eyes and not think about the utter chaos that surrounds me, and most of all, I don't want to think about Brent. We will have to deal with that asshole later.

I'm not sure how much time has passed when I open my eyes. The room is cleared of all the passengers from earlier and Mary is gently undoing the IV on my arm.

"Hold still, this will be quick," she says, sliding the IV out of my arm and covering it with a small gauze. When she is finished, I look over to Kelly and Sara again. Kelly is still a little out of it.

I watch as Sara leans in and whispers to Kelly and gently rubs her arm. Kelly looks lost. Her eyes keep darting around the room, like she is trying to figure out where she is. I realize I've seen this look on her before. It was the day she picked up the fork and stabbed John in the neck.

I stand and begin walking toward Kelly and Sara. As I get closer, Sara begins shaking her head and pointing in the opposite direction. I pause and look around to see what she is pointing at, but I don't see anything. My thoughts drift to my soon-to-be baby, and I decide I have a right to know what is happening. I start walking up to them again, and Sara puts her hand up again. Kelly must notice Sara's movement because she turns around and is now facing me. I watch her eyes go wide, as if I am an animal about to attack her. Only this time she doesn't run away from me. She rises from the floor. Sara tries to pull her back as she stands but fails. Kelly grabs the bloody weapon lying on the ground and rushes at me. Startled, I stop as she quickly closes the distance between us.

"Kelly... babe? What are you doing?" She doesn't answer or stop.

"Kelly, stop!" Sara shouts, jumping to her feet. Kelly doesn't listen to Sara and continues charging toward me. When she is only a few feet from me, Sara shouts loudly, "Sophie, stop now!" *Wait, who the hell is Sophie?*

Kelly halts, then spins around to face Sara. She stares at Sara, breathing heavily for a few seconds before she drops the weapon.

"Sophie, come back here. Tyler isn't going to hurt you. I promise. You know you can trust me. Please, Sophie, just come back over here," says Sara, breaking the silence.

I stare at them both, utterly confused. Why did she call her Sophie? Kelly stands unmoving, staring at Sara. Slowly, Sara walks over and wraps her arms around Kelly, or Sophie, whoever she really is.

"Come on, let's go back to our room. I'll keep you safe," she says, taking Kelly by the hand. As they both head toward the door, Sara calls over her shoulder, "I'll explain everything later, I promise."

The room is quiet. No one speaks, but from the looks on everyone's faces, I'd bet anything we all have the same question running through our minds—Who is Sophie?

The only person I think may know the answer is Chris. "Who is Sophie?" I ask him.

Chris stares blankly at me.

"No idea. All I know is that Sara and Kelly have had a very strange relationship since I met them. You'll need to ask them, not me."

About an hour later, while we are trying to clean up all the mess and still tending to the wounded, Sara reenters the room with Kelly. Tears stream down Kelly's face, and she looks embarrassed.

Sara turns to Kelly. "It's okay, sweetie. It's time to tell them." Kelly doesn't respond; instead, she keeps crying and walks toward the center of the room. I remain where I am—uncertain whether to approach or stay back. Julie doesn't speak or move either. Mary steps aside, giving Kelly a place to sit at the nearby table. Theo's still unconscious, so we moved him to a patient room in the medical center where he can rest and regain his strength away from people bothering him. I glance around the room at the remaining passengers helping with the cleaning. "Thank you for all your help. Can you give us the room, please? Go find some room to get some rest, and we can all meet up later this afternoon to discuss what is going to happen moving forward."

A few minutes pass and the room is cleared out, leaving only Kelly, Sara, Julie, Mary, Chris, and me. My eyes are locked on Kelly and I watch her nervously look around the room at the blood covering the walls, then to the bodies that are outside to be tossed overboard. She shifts her attention back to me. Her face is red and puffy from crying.

"If I'm going to tell you anything, I need to do it somewhere else. I can't be in here with all of this around me," she says.

"We need someone to stay with Theo in case he wakes up," says Mary.

"I'll do it," Julie responds eagerly. I look back and see a flash of a look on Mary's face that I can't quite make out before she shakes her head.

"No. I think you need to hear whatever Kelly and Sara are about to tell us," Mary replies.

Before Julie can respond, Chris volunteers, "I can stay with Theo; Sara can fill me in later. I really don't mind."

"Thank you," Julie responds. She opens her mouth to say something else but then she just bites her lips and nods.

"Okay, so where are we going? The entire ship is a disaster," I say.

"We can go back to Paradise Cove. It's small and stuffy but

it's not covered in bodies or blood," replies Sara. "I know Paradise Cove, and I feel safe there for now. I don't want to be somewhere anyone could sneak up on us."

"Sounds like a good idea, let's go," I say, and we all make our way out the door.

29

KELLY/SOPHIE

I am horrified by the condition of Paradise Cove. The room makes me gag; it smells horrid, like urine and stale body odor. The place is a complete disaster and I almost tell everyone to turn around and go back to the medical center, but I think better of it. I'd rather smell the stink than see all that blood and a pile of bodies that are going to be fish food.

Light from the sun shines through a small porthole in the back of the room, casting only a little light. Sara walks directly to the back of the room and sits at the table closest to the window. Everyone follows. I grab Tyler's hand and we make our way to join her at the table. As we walk, I lean over and whisper in Tyler's ear, "I'm sorry." He looks down at me, then kisses my forehead.

We all take a seat at the table, and once we are all settled, Sara speaks first.

"You've got this," says Sara as she reaches for my hand. I feel like I'm going to throw up. What I'm about to tell these people is something I never share and I'd give anything not to right now, but Tyler needs to know. I look up, and all eyes are on me. I wipe the tears from my face and take a deep breath.

I close my eyes and lower my head. I feel a touch on my shoulder and look up to see Tyler. I smile at him, even though I'm terrified he won't want me after I reveal my deepest secret.

"Babe, whatever it is, we will get through it together. I love you," he says to me.

I stare at him. "Thank you, and I love you too. That's why I need to tell you the truth. This is a secret I've had buried for a long time," I say.

"We are all in this together," says Mary. My attention is drawn to Julie's when her head turns toward Mary, giving her a squinted, confused look. I don't understand what the issue is between them.

I stand from the table and walk to the porthole window, staring out at the vast open ocean. The whitecaps glistening in the morning sun ease my mind. Not wanting to look anyone directly in the eye, I keep facing the water and slowly let my thoughts drift to a time I've tried so hard to forget.

"I was eight years old when my mom first met the man who would become my stepfather. His name was Dave. He was one of the most manipulative people I had ever met. At first, he was so nice and charming, treating my mom like a queen. He put her on a pedestal and she loved every minute of it. She left my father a few months before I turned seven—she caught him with another woman. To this day, I haven't heard from him." I pause and steady my trembling hands. "My mom was grieving at first, as any woman would after catching her husband cheating. Dave saw my mother's emotional state and used that to his advantage. At first, he was nice to me, so when he asked my mom to marry him, I was excited. I even asked if it was okay if I called him Dad. But that all changed shortly after the wedding.

"Dave became mean and more controlling, telling my mom what she could and couldn't do and who she could and couldn't hang out with. She wasn't even allowed to hang around or talk to any man unless he was with her, including her girlfriends' husbands. I only knew all this because I could hear him yelling at her through the walls of my bedroom throughout the night. My mom started staying home more and more and her friends stopped coming over. All except one woman—Sophie. She was my mother's best friend." I let out a small laugh and smile, remembering her.

"Sophie and my mom had known each other since they were kids; growing up, they did everything together. Sophie approved of Dave at the start of their relationship. Dave could charm anyone he met. He charmed Sophie, and she thought he was the best thing that had ever happened to my mom. She was even my mom's maid of honor.

"Sophie's vision of Dave quickly changed after the wedding as well. Sophie noticed my mom had changed, becoming someone she had never been. Whenever Sophie questioned anything, my mom would always brush it off with every excuse in the book. It wasn't until Sophie stopped by our house to surprise us that she realized how bad it had really gotten. My mom had a black eye and swore it was from walking into a cabinet. Sophie knew my mom was lying to her. Sophie was protective of both of us; she refused to leave that night until Dave came home." I glance back to see if everyone is still listening. Then I turn back to look out the window.

"My mom begged Sophie not to say anything, but Sophie was not having any of it. Dave came home from work that night, and I was in my room. Sophie and my mom were in the living room. As soon as he walked in the door, Sophie started yelling at him about my mom's black eye. He said that he didn't give her the black eye and that he loved my mom, saying he would never lay a finger on her. I was terrified, but hearing Sophie defend my mom the way she did made me feel so alive and happy that my mom had someone like Sophie by her side, standing up for her.

"After that night, I didn't see Sophie for a while, and I believed things were getting better. My mom still stayed quiet and did whatever Dave told her, but the yelling and screaming stopped for a couple of weeks. That was until Sophie showed up again unannounced. Dave wasn't having it. He told Sophie that she was no longer welcome in his house. Sophie knew my mom wouldn't argue and was too scared to stop him. Sophie must have known that she would make things worse for my mom if she intervened, so that night she agreed to his terms. She only came over when he wasn't home. I don't have many memories of my mom anymore, other than the times Sophie was over, and I think that's because that was when my mom was at her happiest. This went on for years, and my mom continued to do exactly what Dave said, when he said it."

I turn back to look at the group at the table, all eyes still fixed on me. I glance at Tyler and notice he is listening intently. Then I shift my gaze to Sara, who is fidgeting with her hands, looking down. Sara knows how difficult it is for me to share this story. I wipe the tears from my face and rub my belly, thinking about my unborn child. I walk from the window to the table and take a seat next to Tyler.

"Take all the time you need, babe. We are all here for you," he says as he rubs my back.

I smile at him, then continue.

"Around the time I was thirteen, Sophie started getting sick; she was diagnosed with a terminal illness and had stopped coming over entirely. My mom did her best to be by Sophie's side as much as possible. Other than an aunt that we didn't really speak to much, Sophie was all we had. Dave got angry at my mom because she was spending so much time with Sophie at the hospital, and started telling her she needed to stop going there so much. But Mom started standing up for herself. She told Dave that if he had a problem with her visiting Sophie, he could pack his things and leave. I expected him to hit her or, at the very least, start screaming. But he never did. He agreed to her seeing Sophie. I think, deep down, he knew Sophie was going to die soon and that she was all my mom had left.

"A few months later, Sophie passed away and my mom was a wreck. Something inside her had broken, and she no longer cared about anything but me." I take a breath and sigh. *I can do this, just tell them the truth.* "She told me she was sorry that she let Dave's behavior take hold of our lives for so long, and she was going to figure out a way to leave him and get us out of that life and the tiny apartment we were living in. She promised me all kinds of spectacular things. The type of things that made me dream of a new life. Later that night, my mom cooked the most extravagant dinner she'd ever made us. The table looked as though it were a Christmas dinner straight out of a magazine." I take another deep breath, doing my best to control my emotions, knowing the next part is something I've tried to hide my entire life.

Sara reaches over and rubs my hand. "Everything is going to be okay," she says.

I exhale and do my best to relax. I don't get up and walk away, even though everything in my body is screaming to run. "The food looked and smelled amazing. She'd made everything you'd expect on a holiday: ham, mac and cheese, green beans, stuffing. And my favorite– homemade mashed potatoes with tons of butter. I was starving, eager to dig in. My mom stopped me, saying we had to wait for Dave. I rolled my eyes. I hated that everything revolved around Dave.

"Not too long later, Dave finally walked in the door, filthy with grease and smelling like he'd been drinking all the way home, as usual. We both knew not to say anything, having been used to his abuse and drinking for years. She went to kiss him, but he pushed her away and demanded that she get him a beer.

My mom didn't argue but I remember watching her smirk as she walked away, like she had something planned. I watched her slowly walk to the fridge, grab Dave's beer, then set it down in front of him at the dinner table. He'd already started to eat. My mom sat right beside him, as she always had, and I sat as far away from him as possible, as usual.

"Together we ate in silence, well, other than Dave's loud ass chewing and moaning. I always hated the way he ate; it reminded me of a cow. When he was finished, he stood up and left, as usual, leaving his mess for one of us to clean up. But this time, my mom stopped him by gently grabbing his hand. He yanked it away and called her vulgar names, telling her not to touch him again. She flinched a little but then she calmly told him she had dessert. She left the table and fetched two plates of cookies. One was chocolate chip, Dave's favorite, and the other was snickerdoodle, mine and my mom's favorite. Dave ate the entire plate of chocolate chip cookies, which was expected. When he finished, my mom smiled and asked how they were. He didn't answer, and I noticed my mom still had that smirk on her face."

I bring my hand to my face and cover my eyes. My heart is beating rapidly, knowing where this story is going. "My mom then stood from the table and started picking everything up. She stopped in front of him and told him she wanted a divorce, and she and I were leaving. He reached up and slapped her across the face so hard that it left a full handprint. She didn't say anything; instead, she continued cleaning up the dishes. Once she had an entire stack of plates in her hand, she looked down at the plate that had his cookies and then looked him dead in the eye and said, "I'm so happy you ate all of those cookies," and smiled. Standing in the middle of the room, she slammed the plates to the ground. I was shocked.

"She also told him that if he laid a finger on her or me again, she would call the police. Dave's face turned bright red. I'd never seen him that mad before. I was petrified about what he would do next. My mom was terrified too, because I could see her hands trembling, but she refused to let him have the power any longer. She stood her ground, still with a smile on her face.

"Dave grabbed a plate from the table and threw it as hard as he could at my mom. It hit my mom in the face and then shattered against the floor. Blood instantly began to pour from her face— it'd knocked out a tooth and split open a deep gash above her eyebrow. My mom dropped to the floor, grabbing her face. She didn't scream until she looked at the blood on her hands. Dave ran to her and lifted her off the ground by the neck with one hand, telling her that if he couldn't have her, then no one could. He then went on telling her that no one would ever want her, that she was a nobody and a whore. She was an arm's reach away from a glass candle holder; she reached for it as he choked the life out of her. She used all her strength and grabbed it, then she swung it as hard as she could against the side of his skull.

"He let go and started screaming. His head was bleeding so badly. I remember watching my mom crawl away as fast as she could, screaming at me to run. But I couldn't move. Bloody handprints covered the front door by the time she was able to get up and open it. As soon as she opened the door, Dave jumped up and ran toward her, screaming for her to come back. I remember him grabbing her by the hair, stopping her. She turned around and let out an ear-piercing scream.

"The last thing I remember Dave saying is, 'I told you... you... are...nobody. If I can't have you, no one can.' Then he shoved her, and she flipped backward over the railing. She fell three stories down onto the concrete. Everything that happened afterward was explained to me by my psychiatrist."

I take a moment to catch my breath before I finish the story.

Sara reaches for my hand again, and I look up at her. "Want me to finish it? I know everything from here," she says.

"No, thank you. I need to do it," I say, wiping a tear from my cheek. "After that night, I don't remember anything until six months later. I remember waking up in a hospital and asking where I was. I later found out it was a psychiatric care facility."

30

KELLY/SOPHIE

"They told me I'd been admitted six months earlier and that I had told everyone my name was Sophie. They believed that I had a mental breakdown after witnessing my mother's death, that my psyche had broken, and my mind created a second personality named Sophie. The medical term for it was *Dissociative Identity Disorder*. I was questioned about the events that had happened— for me, it felt like the night before, so I remembered a lot of details. That's when the doctors explained to me that seeing my stepfather murder my mom and then attack me was so traumatic that it must have caused my mind to break, creating the alter.

"Apparently, I stabbed Dave multiple times in the neck with a fork. But if I hadn't stabbed him, he would have died within minutes anyway because the autopsy showed that he had been poisoned. I'm not sure if they found out where the poison came from, but I knew. I knew my mom did it with the cookies.

"I stayed at that psychiatric facility for the next year, learning about DID and what triggers my alter, Sophie, to emerge. A big trigger for me is stress or some kind of trauma. Sophie has been hiding deep in my mind since I was a little girl.

I thought she was gone, but that all changed when I learned that *she* killed John. That's also why I can't remember anything that happened that night either. But I remember sitting in the Stardust Lounge right away after the incident. Sara is the only one who knows about Sophie, and she's the only one who knows how to make her go away when she comes out. The last time Sophie came out was when we were in the medical center and I looked down and saw myself bleeding. I remember thinking that my baby was going to die. Then the rest is darkness. I'm grateful Sara was there to help. I'm so sorry for not telling you."

"It's okay, babe—I get it. That's a lot," says Tyler, rubbing circles on my back. *Thank you for not hating me.*

Everyone is silent, staring at me. I look away to the window, embarrassed and ashamed.

"So, how do you and Sara know each other?" asks Julie, breaking the silence. Thankfully, Sara answers.

"I met Kelly—actually, at the time I met 'Sophie'," Sara says. I shift my attention back to the table to listen. "At the facility. My mother was also a patient there. My mom also had a personality disorder, and her alternate personality was extremely depressed and kept trying to commit suicide. Anyway, while I was visiting my mom one day, I met Sophie. She was playing a board game with other patients, and she was the only other person in the room who looked to be around my age. I asked the nurses if I could play with her, and they said it was okay.

"The first thing 'Sophie' told me was that she was there for killing her best friend's husband after he had murdered her. She then told me that her husband would be back at any moment to pick her up and take her home. Sophie fascinated me and the doctors all thought it was a good idea that having someone my age, the same age as Kelly, could potentially help... how do I say this, umm... bring Kelly back to her own self." Sara looks over to me, making sure I'm doing okay. I nod, and she continues.

"Eventually, something worked, and Kelly came back. At that point, I almost had to start over again, because 'Sophie' knew me, but Kelly didn't. Anyway, we've been friends ever since. After Kelly was released from the facility, she moved in with her aunt, who didn't live too far from where I lived. To this day, Kelly is my closest friend. She was there for me when my mom passed, and I've always been there when she needed me, and that will never change." Sara walks over to me and gives me a tight hug, and it takes everything in me not to start crying again.

"Thank you, Kelly, for trusting us with something so personal, and thank you, Sara, for being there for her," says Mary, pulling me from my thoughts.

After they all ask me a few more questions, we all exit Paradise Cove and then head back toward the medical center. It was absolutely terrifying to tell everyone my truth, but I'm glad I did. Now all that's left to do is figure out how the hell to get off this damn ship.

31

TYLER

That was a lot to process. I never would have, even in my wildest dreams, imagined Kelly sharing all of that with me. I don't ask her any questions as we walk back to the medical center; I can only imagine how overwhelmed she must feel after revealing her past. I give her space, and when the time is right, we can talk more about everything when we're alone.

Once we arrive at the medical center, Sara and I make a plan. The day is halfway over, and with only so much daylight left, we ask Julie and Mary to stay with Theo and ask Chris to gather all the passengers so we can figure out our next steps for survival.

Once everyone is in the medical center, I do a headcount, and only thirty-six people are here, a shocking number compared to just months ago. I look around the room at all the faces. Only one person worries me: Brent. I don't trust him at all. The scowl on his face as he stares at me hints at what might be going through his mind. Not letting Brent intimidate me, I turn away and focus my attention on Sara, who is standing in front of everyone.

"Everyone, listen up," she shouts, loud enough to make

everyone stop and face her. "We need to get this ship cleaned so we can make the space livable while we wait for rescue."

"When the hell is rescue coming? If we haven't been rescued by now, we're never getting rescued," says a man from the back. I look to see who said that, and when I realize it's Brent, I roll my eyes. I knew he was going to be an issue.

"Then go jump off the fucking ship and see who gets rescued first. If you are going to complain and cause issues, leave. We don't have enough supplies or energy to waste on convincing you to stay." Sara stares Brent in the eye while speaking directly to him. Brent, wisely, doesn't argue with her and remains silent. I smirk; impressed. She said what I wanted to and seems to be a natural-born leader. My father would have loved her guts.

Now that she has everyone's attention again, she gives direct orders: clean the medical center. Dump the bodies. Bring all the food and water on the ship to the medical center to keep everything in one place. Once she's finished, we all spring into action, each of us going our separate ways.

It doesn't take us long to get the place clean. The first thing we do is remove the dead bodies. I help carry the bodies to the 4th floor at the broken railing. When I step outside onto the deck, the crisp, cool air is refreshing as it blows across my skin. As each day passes, the temperature drops a few degrees, hinting that winter is quickly approaching. A few passengers who helped with disposing of the bodies months ago volunteer again to toss them over. One by one, we toss each body into the water. Before all the bodies are tossed, I head back inside to the medical center and see if anyone else needs help.

"This is all the food that we could find," says Chris, setting down a very small amount of canned food.

"We will have to make it work," I say, unsure of what else to say. Chris just stands there, looking at me as if he expects me to give him an order. I glance around the medical center at everyone doing what they can to clean up the mess, and look back at Chris. "Do you mind keeping an eye on Theo, and telling Julie and Mary to meet me out here?"

"Sure thing."

Chris walks over to Theo, and I wave down Sara and Kelly. Sara, Kelly, Mary, Julie, and I check out the remaining food. We quickly realize our biggest problem is how to ration the small amount of canned food left, since it definitely won't last long. There are only so many ways to split up the food.

While we're trying to figure out a fair way for everyone to eat, Chris rushes over. "Theo is awake!"

Oh, thank god. I look at Julie, expecting her to share my excitement, but I'm surprised to see that she's crying. She lifts her eyes and her gaze meets mine.

"I'm sorry," she says, wiping away the tears. "They're happy tears." I'm not sure what to say. Instead of saying anything, I turn on my heel and head to Theo's room.

Theo looks much better than I expected. He still has a large bandage covering the wound. I can't help but think that it could have been me in that bed instead of him. It's not hard to imagine when I'm staring at him, it's like looking in a mirror. While I'm standing there, staring, he looks up at me and tells me to come over. I stand there unmoving for a moment, then Kelly walks up behind me and gently nudges me.

"Go to him," she whispers in my ear. We both walk over and sit down. Julie and Mary follow shortly after. We chat for a while; I joke and tell Theo he's been in a coma for weeks. He almost has a heart attack before I tell him the truth– that it's only been hours.

While we're talking, one of the guys helping toss the bodies comes running in, screaming about fish. He explains that while they were throwing the bodies overboard, someone spotted a large fish. My first thought is it had to be a shark, since they were throwing bloody bodies into the water. Then he starts talking about catching fish for food. We've tried catching fish before, but we've never used bait before before—maybe that's the missing piece?

"We were just talking about the food supply," I say.

"What about food?" asks Sara, walking in. Before I can answer, Theo speaks up.

"I think it's a great idea, Jordan," he says to the guy. *Jordan, that's his name. See, that's why Theo is a good leader. He remembers the important things, like names.* "But, how are we going to catch any fish without fishing rods or hooks?" He makes a good point.

"Let's ask what everyone thinks. Maybe someone can help come up with an idea," I say.

"That's actually a really good idea. Can you all get everyone together so we can ask now?" asks Sara. Her tone tells me it's not a question but an order.

"Sure," I say, standing up. I look over at Kelly, "Wanna help?"

"Yeah," she answers.

Back in the medical center, the bodies have all been moved but the bloodstains remain. I stand there listening as Sara talks about the fish and asks if anyone has ideas for catching them. Everyone begins shouting different ideas, most of which are impractical or unachievable. I don't know much about fishing; that was always Dad and Theo's thing, so I listen quietly.

Eventually, everyone agrees on one plan: use all the nets designed to cover the swimming pools at night, tie them together, and attach small weights from the gym to the edges to create a casting net. I think about the idea and agree that it's a good one. In theory, the plan will work for larger fish; the holes in the pool netting are too big to catch smaller ones, but I keep this to myself. With food supplies running low and large fish supposedly nearby, anything will help us. The idea of fishing for food sparks hope, which is a feeling I haven't experienced in a long time. Several people volunteer to help make the net and find anything useful. After a few hours, the casting net is ready.

Eagerly, we all rush to the 4th floor. Theo has a hard time getting there, saying he's still a little dizzy, but he doesn't want to miss the first casting. Julie and I help him walk down. When we finally make it to the 4th floor, we look out at the water and spot a fin circling. I can't tell if it's a shark, but I think it is. It must be young because it's not very big. As a team, the people who built the net are the ones who get to cast it. When they throw the first cast, it misses; instead, the net catches one of the floating bodies. No one says anything. Anticipation is high and nerves are on edge—the thought of eating something other than canned food weighs on all of us.

The body is pulled up and tossed overboard again. Everyone looks away because it's too hard to see one of our own like that. Once the body hits the water, we all watch as the large fish swims around. This time, when the net is thrown, it lands on top of the fish. After many minutes of the fish fighting and almost pulling people overboard, someone steps out from behind the group and throws one of the makeshift spears at the fish. The spear pierces directly into the fish, and it stops swimming. I turn around to see who threw the spear, surprised at the precise accuracy. It's Xander.

"What? I used to do spear fishing tournaments with my dad. It was kind of our thing," he says with a smile.

"That's awesome." I respond.

It takes several people to lift it; it must be heavier than I expected. Once it's on board, we discover it's a shark. I can't help but smile, knowing I was right. The shark is large enough to feed everyone for a few days. Karissa volunteers to help gut and prepare it for cooking. A few others speak up and offer to make a fire pit.

As the day ends and night falls, I can't sleep. I know that I could go back to my room if I wanted to but Kelly is here, and she said she feels safe being around Sara, and Theo is staying here too, so Mary can keep an eye on him. I'd rather stay here around everyone because I don't really feel safe alone. There are people who believe I shouldn't have killed Logan. Lying here in one of the medical beds, I stare up at the ceiling. The thought of cooked meat sounds so good. I'm so excited for the food that I can't turn my mind off.

Karissa and Xander work through the night, filleting the shark, and I am so thankful because, once again, there's something dear ole Dad never took the time to teach. Xander, with his bandaged leg, helps Karissa. A few others gather materials to create a fire pit outside on the patio. Someone had the bright idea to take all the wood out of the ship's saunas to cook with. I'm not sure how long I stay staring at the ceiling, but eventually I drift off to sleep. I wake up to the smell of cooked fish filling my nostrils.

The early morning sun is bright through the windows. I'm not the only one who must feel safer here in the medical center, because several blankets are laid out on the floor where others slept. There are still bloodstains throughout, but most have been cleaned. I make my way outside to the firepit, where Karissa and Xander have a tray filled with cooked fish next to the pit. My mouth starts watering the instant I see the pile of cooked meat. Kelly and Sara are sitting off to the side, watching Karissa and Xander finish cooking. I walk over and take a seat next to them.

"Good morning, how did you sleep?" I ask, looking at both of them.

"I slept fine I'm sorry I didn't want to go back to our room,"

Kelly says, reaching out to take my hand. She's wrapped in a sheet, and I desperately want to curl up next to her. I stay put and give her space, even though she isn't pushing me away. "How did you sleep?" she asks.

"Ehh, not that great. I don't know when I finally fell asleep. There is too much on my mind."

"I get that. But on the bright side, we have a hot cooked meal today," says Kelly. One by one, passengers start making their way outside near us. They all look as exhausted as I feel.

"Get everyone together, food is just about ready," says Karissa, slapping the last piece of raw fish down. When she lays the meat on the hot metal, it pops and sizzles. Smoke rises, and my mouth waters so much I have to swallow. I glance around, and we are all desperately waiting to get a taste.

"Make way for Karissa to bring the food in," commands Sara, waving her hands to get people to move. Karissa and Xander walk in and set the food down on the table. Steam is still rising off the hot food, quickly filling the room that just the day before smelled of blood with the scent of cooked, savory fish. Because all the food and supplies were brought up to the medical center, we have plates and utensils to use. We all grab a plate, and Karissa gives each of us a piece. Once we are all sitting down with a plate filled with the flaky white fish, you can hear people devouring their food from around the room. Some are chewing slowly to enjoy the flavor, while others are nearly swallowing the meat whole. I take my time with small bites. Kelly, sitting next to me, is nearly finished with hers. *She's eating for two.* I look down at my plate, then back at hers. I don't need all of it, so I cut it in half and dish it onto her plate.

"What's that for?" she asks, smiling.

"I gotta keep you and the baby healthy, of course."

"No, babe, you eat it. I'll be fine, you eat it."

"Kelly, you—"

"If you're not eating it, then I will," says Brent, grabbing half of the cooked fish and shoving it in his mouth. When his eyes land on mine, I feel an anger build inside me like never before.

"What the fuck do you think you're doing?" I spew. "Spit that out right now."

"Too late," he says, sticking his tongue out. "It's all gone."

Theo and Sara both glance at us from across the room.

"What is going on?" asks Theo. I don't answer him. I look down at the white plate in front of me. There is still a piece of fish on my plate. I slide it off my plate and onto Kelly's. Before she has a chance to say anything, I rise to my feet and slam the plate across Brent's smug face. Instantly, the plate breaks into multiple pieces, falling to the ground. Blood streams from Brent's nose, soaking his lips and chin.

"Tyler, stop!" Kelly screams.

"What the fuck, asshole," murmurs Brent. In the corner of my eye, I see Theo trying to stand, and Julie and Mary both stop him. I look back at Brent, who looks pissed. I don't let him up off the floor where he's sitting. I kick him in the chest as soon as he starts to stand, and he falls back. I jump on top of him and slam my fist into his face over and over again.

"I'll fucking kill you, just like I did that piece of shit Logan, if you ever even look at Kelly again." With each punch, his face becomes bloodier.

"Get off of him," I hear a man shout, and I'm suddenly yanked away from Brent. The whole room erupts into chaos with everyone yelling. I look over to see who pulled me back, and it's Cyrus, another one of the guys who believed Logan didn't deserve to die. Within seconds, there's a loud bang, almost like a gunshot, and everyone falls silent. I turn to see where the sound came from and it's Sara—she's holding two frying pans. *Where the hell did those come from?*

"Enough! We're not doing this again. We don't take food. We're not fighting anymore."

Brent mumbles something from the floor that I can't understand.

"Brent, so help me. If you try something like that again, I'll kill you myself," says Sara through gritted teeth.

"And if she doesn't, I will," Theo says. "From now on, we'll keep fishing for food and eat as much as we can. We shouldn't fight over food like this." I glance at Brent, who has a swollen eye and a bleeding, busted lip. Our eyes lock, and he doesn't say anything. I can already tell this isn't over.

"We should be celebrating having a hot cooked meal. Instead, you want to fight about it and steal off someone else's plate? Look around. We're all that's left. We're in the middle of the damn ocean. There are plenty of fucking fish in the sea. Don't let this happen again, or as Sara and I said, we'll fucking kill you ourselves," Theo spews with anger that makes his forehead vein bulge. "I'm not Logan. I don't want to kill just to kill, but put anyone else in danger, and I won't hesitate to take you out."

Nobody argues or says anything else.

PART 3

JANUARY 2013

EIGHT MONTHS DRIFTING

32

THEO

Winter has officially arrived. I have no idea of the ship's exact location, but I certainly know it's far from our starting point because it's never this cold at home. The water is dotted with floating icebergs, and the entire deck is covered in ice and sometimes snow. The temperature gauge hanging on the pool deck is iced over, and it reads -19, but I'm almost certain it's Celsius, not Fahrenheit. The cold breeze from the ocean is so fierce at times that I'm sure it could freeze almost anything within minutes. Fishing is no longer possible, even with the fishing pole we lucked out and found in one of the crew's belongings, but fortunately, enough fish were caught just before the first freeze to feed everyone for a few weeks. The freezing conditions are keeping the fish frozen on the top deck.

During the initial planning three months ago, when we caught our first fish, we had to come up with other ideas for survival as well. The broken water slide on the top deck was dismantled and repurposed as a rainwater catchment system. We gathered empty trash cans and placed them at the base of each broken slide piece to collect rainwater, which was then transferred into used bottles for storage. However, the collected water supply is now nearly exhausted. Since it's winter and there's no rain, people have begun collecting ice around the deck and bringing it inside in bowls to melt. Unfortunately, that yields much less water than the rainwater we've collected previously.

It took some time, but we were able to establish a working community again. Sara is now my co-leader, and we make a good team. Since October, only a few people have died—three who refused to eat or drink. We all urged them repeatedly not to give up but they were too far gone. I guess everyone processes hopelessness differently.

Julie spends a lot of time in the library, even though the entire space is infested with mold and mildew on the books. I've told her at least a million times to stop going in there, or she's going to get sick. She doesn't listen and does as she pleases. She keeps to herself and only associates with me when necessary or during decision-making. Each night, she falls asleep in the living room area of our suite on the sofa, reading a book. She remains distant from everyone, claiming she is reading and prefers solitude. She has become a shell of herself, and I'm not sure how to help her.

Rations are limited to two small meals each day. Everyone gathers to eat together at sunrise and sunset. Meals are no longer cooked outside on the patio where the fire pit was initially used because it's too cold. Instead, we thaw the fish as best as we can and eat it raw. The doors stay closed as much as possible to keep any heat inside. As temperatures drop, more people are falling ill and struggling to stay warm. We've tried countless times to build a fire pit inside but nothing ever works. Each new area we try, we face the same problem—we can't get the smoke out without leaving the doors fully open. It's too cold for that. The only option is to leave the fire pit outside the medical center.

Fortunately, the ship has stayed quiet and uneventful for the past three months. I try my best to keep everyone's spirits high, but with the fish running out soon and the temperature dropping each day, the idea of rescue is starting to sound more like a fairytale.

33

JULIE

FEBRUARY 2013

The early morning sun filters through the balcony window, warming my face, and it feels good. My eyes flutter open, and I let out a soft yawn before stretching. Even with multiple layers on, I feel the frigid winter air around me. Quietly, I slip out from under the warm blanket, grab one of the sheets we have layered on it, wrap it around my body, and make my way to the glass door to enjoy a little more of the sun's warmth. The balcony is covered in snow, which has risen at least three feet up the door. I let out a breath and watch as it fogs the window. I feel the urge to reach up and draw something, but I don't. Instead, I turn back, looking toward the mattress, and see the imprint of where I was just lying. Theo is sleeping peacefully, all bundled up under the duvet. My mind races with thoughts of the past, and I can't help but think about what the future could look like.

I look around the large room, which feels more like a prison cell than a luxury cruise room anymore, and I recall the day I first entered it, seeing the king-size bed perfectly made, with chocolates on the pillows and rose petals scattered across the perfectly white duvet. Oh, what I would do for a piece of chocolate right now. I remember stepping into the oversized bathroom of the suite that Theo specifically selected for us and gasping when I saw the walk-in shower for the first time. The thought of hot water flowing over my skin makes me realize how cold I am right now. I pull the blanket a little tighter around me.

A loud thump pulls my attention, and I realize it's a piece of ice falling on the patio. I look over to see Theo still sound asleep, and my eyes drift to the nightstand where my book, *The Last Song,* rests. Instantly, my body stiffens. Cautious not to wake Theo, I tiptoe to my nightstand and grab the book. I pry it open and release a breath of relief as the handwritten letter comes into view. The papers, now fragile from being handled so many times, are nearly torn at the crease. Holding the letter in front of me, I stare at the words, which now seem like they were written a lifetime ago. I peek over the edge and see Theo's silhouette curled up, facing the opposite direction. I realize so much has changed since the day we boarded the ship and I began writing this letter. Initially, I never planned to give it to him. Thinking back on all that's happened over the past months, I realize that our relationship is no longer what it was. I'm torn on what I should do. Do I give him the letter, or do I forget it all and move on? So much has changed. For the hundredth time, I reread the letter.

Dear Theo,

You've been a part of my life for a long time, and I know you better than I know myself, which is terrifying. You're incredibly brave and always put others first. You see the good in everyone, even when no one else does. You've loved me through all our highs and lows. You were there when the doctors told me I could never have children. I've never told you the full truth about why I can't have children, or that even before that appointment, I already knew it wasn't possible for me. I'm sharing this through this letter because I am unable to do so in person. There are many things about our past that you don't know and were never meant to find out. There are things about your father that he never wanted you to know, and he paid a lot of money to keep those secrets hidden. Over the years, I've watched you work so hard to impress your father and become the person he always wanted you to be, and that hurts me because you are nothing like that cruel man. His money and power turned him into a wicked man, with more secrets than you can imagine. It's not my place to reveal everything that I've learned about his past, and I know you loved him more than anyone, so I don't want to ruin that for you. But I promised myself—

The sound of a deep yawn echoes through the room, startling me. I look up and see Theo shuffling and beginning to roll over. Quickly, I fold the letter into the book. Theo turns over, and his eyes are staring directly at me. He squints in the blinding sun. I step in front of the sun to help block out the light, and he smiles at me.

"Good morning. What are you up to?" he asks hoarsely.

"Not much. I couldn't sleep, so I was reading," I answer softly.

My voice is gentle yet nearly cracking as I speak, with emotions swirling inside me. I feel a tear escape and roll down my cheek. I walk over and sit on the bed beside him. Our eyes meet, and he gently reaches out to grab my hand. I quickly pull away to wipe the tear from my face.

"I'm fine." I glance back at the icy balcony. "It feels like we're never getting off this damn ship, and I'm freezing."

Theo lifts the blanket and scoots back on the bed a few inches, making room for me. "Come under the blanket; I'll warm you." I don't move right away. "You can't think like that, honey. Let's stay positive and believe that rescue is coming; it's only a matter of time," he says, softly. I stare at him blankly for a moment. So many emotions run through me. There's no way he truly believes that anymore.

"How can you just lie there, pretending everything is fine? We've been out here for what—eight, nine months? We should have been rescued by now. People are dying all around us. So many people have literally jumped, just giving up. We are down to what, thirteen survivors left on this ship—and that was the count last night. Who knows how many more have jumped in the middle of the night? I can't do this anymore, Theo. This isn't living. We need to find a way to get off this ship, or I might jump too."

I'm angry, sad, and regretful all at once. I close my eyes, take a deep breath, and exhale, relieved to have let my feelings out. When I open my eyes, Theo doesn't look relieved at all. He looks at me, but says nothing. His eyes glisten, as if he is about to start crying. Then a moment later, he does. He slumps over, grabs his face with his hands and loses it. He sucks in a deep breath and sobs. Deep, wracking sobs that make my heart ache. This is a side of my husband that I never see. And honestly, I don't know what to do.

"Julie, you're right. But if I don't keep pushing forward, no one else will. I'm responsible for everyone, trying to keep everything together so we don't end up like we were a few months ago. And it's fucking killing me and no one cares." He sniffs and wipes his nose with the back of his sleeve. "Nobody gives a shit about how I'm dealing with all this, and if they did, it wouldn't matter because everyone's looking to me to keep my shit together so the rest of us don't end up dead." He goes quiet for a moment, sitting there.

Then something changes in him. He looks up at me with cold eyes. "You know what? I'm so fucking tired of these people. They're ungrateful, and I'm sick of all their shit. We *should* just jump. Hell, no one's coming to sa– Fuck!"

I stand in place, stunned, unable to speak, my thoughts running wild. I stare at Theo, and for the first time in our relationship, I see he's truly at his breaking point. I think back on the past months and realize he's right—no one has ever asked how he's holding up mentally. At least I haven't. The letter I wrote flashes in my mind, with all the untold truths it contains. I can't help but think it would be so much easier at this point if we did make the jump.

Then I think about my mother, the car accident, and my deadbeat father, and I absently rub my fingers over the thick scar on my abdomen. I feel a sudden twinge of anger that pulls me out of the self-sabotaging hole I'm in. *This is not how it ends for me.*

"No, Theo, this is not how it ends for either of us," I say, reaching for his hand. He lifts his head and looks at me, his eyes shiny and red. I don't know what to say. I let go, feeling wrecked inside, before standing up. "I need to go for a walk to clear my head. I'll meet you all later."

"Are you kidding—where are you going? You're going to freeze if you go outside," he tells me, wiping his face with his sleeve again.

"I'll be fine."

"Be smart and come back when you get cold. I'll go find Tyler and Kelly—they're probably in the medical center. Come there when you're ready."

"Sounds like a plan," I say, trying to keep my voice steady. I turn away and walk out the door without saying goodbye. The moment it closes behind me, the emotional wall I've built collapses and all my feelings hit me like a train. I can't breathe from crying so hard. I need to find a place to be alone.

34

THEO

Walking into the medical center, I immediately notice that all the remaining passengers are here, except Julie. In the back of the room, Sara and Chris are arguing with Brent about something I can't quite hear. Brent has become everyone's biggest problem over the past few months. He is constantly trying to control people and tell everyone what they should and how they should be doing things. Because of his constant arguing, Sara and I have assigned him the worst job on the ship: emptying the bathroom buckets and fetching fresh water from the ocean—an especially tough task when temperatures are below freezing. Given his attitude toward everyone, no one objects, nor does anyone want to be stuck with his job. Brent's other job is to collect the fish from the top deck daily for our meals. Rations have become so scarce that meals are limited to once a day. As a result, most passengers have lost a lot of weight and are weak. Mary keeps saying that soon, passengers will start dying if we don't find more food, but unfortunately, there's nothing anyone can do about it. I walk over to Kelly and Tyler and check on their well-being. They both say everything is fine, and for Kelly, that may be true, but for my brother, I know it's not. Kelly looks healthier than most of us because she's been getting a little extra each day at mealtime. Tyler, on the other hand, looks much worse than almost everyone else. Whenever I bring up concerns about his weight, he always gives the same explanation: that Kelly and the baby need the food more than he does. I argue that it was fine to give his food away when everyone was getting two meals, but not now that we all only get one. I try to make him understand that if he keeps giving away his food, he is going to die, and then his baby won't have a father at all. This doesn't change his mind. Tyler is one of the most stubborn people I know.

I take a deep breath and think about what to do. Maybe I can see if everyone wouldn't mind giving Kelly just a bit more of their rations. I spoke to Mary, and she said she doesn't need much, maybe a few hundred extra calories, so we could–

Brent's yell from across the room tears me from my thoughts. Everyone else turns to see what all the commotion is about. He's shouting directly in Sara's face, his finger is pointed at her, and he is calling her names and saying other things I can't make out. Chris jumps in to defend her and Sara stops him just as fast, but as soon as Sara turns back around to face Brent, he swings a fist. There is an audible crunch as she falls backward, and blood pours down her face.

Chris jumps up to defend Sara once more, but he's no match for Brent's stocky build. Brent picks Chris up by the neck, slamming him to the ground, and begins pounding on his face. One hit after the other, blood flies through the air with each swing. Screams echo around the room at the sight. Tyler and I both jump up and rush to stop Brent. Tyler is the first to reach him and tackles him off Chris. Brent's head slams against the glass doors so hard I think the glass shatters.

I jump in and throw Tyler back. Brent is now lying on the ground and scrambling to escape. I straddle him, pinning his body down. Before Brent has a chance to attack me, I slam my fist into Brent's eye. I feel the skin on my knuckles split, and my hand becomes bloody. Brent's eye instantly swells. All the shit Brent has said to me over the past couple of months rushes to the front of my memory. *You will never be as good a leader as Logan. You're weak. You should have been the one to die, not Logan. You're never getting off this ship alive.* With each memory, I get angrier and angrier.

Then I think of what he said to Tyler. *You'd better watch your back. I'm coming for you.* Everything I've been keeping bottled up pushes its way to the surface. *Be a man, don't let anyone belittle you. Take control. Make me proud.* That was my father's mantra. His voice in my head is a constant reminder to be better. To be stronger.

Then there is Julie. She is never by my side anymore. I feel so alone. Fuck it. My father was right. I will not let these people walk all over me. I am the leader. I am stronger. I grab Brent's skull with both hands and slam it against the hardwood floor. With each forceful slam, blood splatters across the floor. Mary screams, telling me to stop, but I ignore her and continue. Rage floods me.

Brent begins to fight back, bucking his hips up and down, causing me to lose my balance and bounce off him. While I'm on the ground, Brent takes his foot and bashes it into my face. The pain is instant and excruciating, but it fuels my rage even more. Before I can stop him, Brent crawls away and takes off running. I jump up, almost losing my balance, my vision blurry as blood drips into my eyes. Full of anger, I chase after him down the hallway. When Brent comes into view, I watch as he opens the door to the top deck, and cold wind fills the hallway, covering the floor with powdered snow. Brent runs out the door, leaving a vivid red trail of blood across the powdery white snow. I follow the trail, leading me directly to him.

I do my best to speed up and close the gap between us. The fierce wind and freezing temperature are so intense that my exposed skin immediately begins to freeze, sending a shockwave of pain through my body. My face is pelted by icy snow as the wind howls around me. Just as I get closer to Brent, I see him reach for the fish basket and start digging at the snow that is blanketing the fish. Now mere yards away, I realize exactly what he's doing.

I push myself to run even faster. I slip on the ice, falling flat on my face and I feel my cold, frozen skin split open. Crimson stains the white snow even more. I lift my head to look at Brent and our eyes meet. Wind whips the snow through the air, making it even harder to see each other. A second later, the wind slows and everything becomes eerily silent.

Brent's voice booms through the air. "You're all going to die out here. I'll make sure of it."

Suddenly, he lifts the basket high above his head. He holds it there for a moment, then our eyes meet again, and then he hurls the fish-filled basket over the railing. I jump to my feet and rush toward the side of the ship, hoping the basket didn't land in the ocean. When I reach the railing, defeat consumes me as I stare at the basket sinking under the waves.

The realization that our food supply is gone hits me and I slump to the ground. Anger boils inside me as I think of all the problems Brent has caused us over the past few months. I ignored all the red flags, and I still gave him a chance to prove himself. I knew that all along, Brent believed everything Logan had ever told him, including that no one would survive on this ship.

Quickly, I stand and take off toward Brent, who is sitting on the icy deck. When I reach him, I slam my fist as hard as I can into his face, splitting his lip and causing more blood to spill out. I continue beating Brent, each punch splattering blood and spit all over the ground. As I start to tire out, Brent opens his swollen eyes and meets mine. He grins as more blood pours down his face.

"Go ahead, kill me, put me out of my misery." He spits a mouthful of blood in my face. His body goes limp under me and he doesn't even try to fight back.

He wants me to kill him, which would be so easy. But Brent doesn't deserve an easy way out. I grip his feet and begin dragging him through the snow. Brent's body feels like dead weight as I pull him toward the door. He groans in pain, begging me to kill him, but I ignore him. I hear the door creak open, and when I look up, I see Tyler there.

"Grab some rope and follow me," I shout over the howling wind.

"Why, what are you doing?" asks Tyler.

"Just get me some fucking rope," I demand, adjusting my grip on Brent.

I watch Tyler take off running inside. A few minutes pass, and Brent keeps calling me names. I stop and kick him in the face to shut him up, and he spits out a tooth.

"Is that all you've got, you little bitch?" he asks, spitting blood. I know what he's doing, but fuck that. I won't give him the satisfaction. Within minutes, Tyler runs out the door with the casting net in his hand. He throws it on the ground beside me and kneels. We both begin undoing the rope connected to the net and after a few minutes, we have him tied up, and I drag him inside. The watching group steps aside and lets us pass. As we walk, someone in the group begins shouting, "Kill him!" as we descend the stairwell. I slow down a bit, feeling exhausted, almost too tired to keep going. I look back at Brent, and our eyes meet. He grins again. Anger fills me once more, and adrenaline flows through me. I pull Brent down the stairs and with each step, Brent's head bounces like a ball.

"Where are you taking him?" Tyler asks, a confused look on his face.

"Stop asking questions and just follow me," I say.

Tyler doesn't protest and does what I tell him. No one else in the group questions what's happening. I stop when we reach the 6th floor and look over the railing at the atrium.

"Tyler, give me the rope and go down to the center of the atrium," I say.

Without hesitation, Tyler hands me the rope and takes off. Tyler quickly runs down two floors and through the only open entrance to the atrium floor. I grab a piece of broken marble from a table and tie it to the end of the rope. I look over the edge and see Tyler standing in the center of the room, staring up at me.

"Tyler, watch out! I don't want you to get hit." I shout, then wait for my brother to step aside. Once he has moved out of the way, I toss the rope to the chandelier—I miss. I shout in anger and try again; this time, the marble piece hits the ring on the chandelier and falls through the center. I let the rope go, and the marble's weight brings it down to where Tyler is standing.

"Throw it up here, Tyler," I shout. He is weak but does his best. He stands there swinging the rope with the marble attached like a fan blade and then lets go. The rope flies up and I miss grabbing it by just a few feet.

"Try again Tyler, you've got this." I say.

He winds the rope up once more, using all his strength and then lets go, this time I'm able to catch it. I set the marble end aside and turn to face everyone. "Listen up, everyone. I want all of you to remember what happens when you turn on one another. Brent is going to be a constant reminder." I look around the room for something to help me make my point. My eyes land on a white lifebuoy out of place in the corner.

"Bring me that lifebuoy," I demand.

Xander is closest and reaches for it. I watch as he bends over to grab it and see him clutch the spot on his leg that was pierced by the spear months ago. It was infected for a while; he could hardly walk. Now it's finally healed enough to move around, but from the look on his face, I can tell it still hurts. Luckily, the wound on my shoulder didn't get infected and healed quickly. Xander reminds me of myself, and because of that, I know that if he's like me, he's too stubborn to sit around and heal. Xander walks over and hands me the lifebuoy. Then I lift Brent's head off the ground and drape it around his neck like an oversized necklace. I grab the end of the rope and make a noose. The other end is tied to the wooden railing, and I tug on it to check if it's secure. Then I lean over and see Tyler staring at me.

"Come up here," I shout. My brother needs to be up here for this, so I stand there and wait. Brent has also terrorized Tyler for far too long. I look down and kick Brent again to make sure he is still alive. I'm not letting him die that easily.

"Are you sure about this? We can find another way," says Sara.

"Yeah, Theo, you don't have to do this," Mary adds. *Take control. Show them you're in control.* My father's voice echoes in my mind again.

"This is the only way. It has to be done. Brent doesn't care about any of us. He wants us all dead."

35

TYLER

The cold air sends a shiver down my spine as I race up the stairs. When I reach the 6th floor, I hear something above me. What the hell? No one's supposed to be up there. I look up, unsure of what it is, since the floor above is unfinished and blocked off. Everyone on the ship except Julie is gathered with Theo, and he is about to do something crazy. As I head over to where they are, I hear the sound again. I start back up the stairs and look around as I reach the 7th floor. To my surprise, I see Julie rearranging a piece of wood over a hole Theo created months ago. She slides a plank of wood around again, then startles when she turns and sees me.

"What are you doing, Julie?"

"Don't sneak up on someone like that," she demands. "I was fixing this wood. It fell when I walked by it a second ago."

"Why are you walking up and down the stairs?"

"I needed to get some air," she answers.

"Yeah, but–"

"What's going on down there? Why is Theo yelling?" she asks, cutting me off.

"Brent attacked Sara, and Theo is taking it into his own hands to fix the problem."

"What do you mean?"

"I don't know," I say, as I walk past her towards the stairs.

Lost in thought, I suddenly snap back to reality when I hear Theo yelling. "Tyler, where are you?"

The 6th floor is crowded as all the remaining passengers stand back to watch what Theo is about to do with Brent.

"Let this be a reminder to you all. I said we are family. If you put someone else in danger, there will be consequences. We don't have enough resources or energy for any of you to be going against each other. Brent is a traitor to all of us. We're supposed to be protecting each other, but Brent here..." He kicks Brent with the tip of his shoe. "Go ahead and tell them what you did."

Brent grunts but doesn't answer. I'm startled when Theo suddenly kicks him in the stomach full force. "Tell them what you did, asshole."

"Fuck you, and fuck all of them," Brent growls, blood and saliva spewing as he forces the words out. Theo slowly kneels, bringing himself face-to-face with Brent.

"Is that all you have to say for yourself?" Theo asks with a look in his eyes I haven't seen since I was a teenager, when he protected me from a school bully.

"Like I told you before, Logan should have killed you when he had the chance," says Brent, then he headbutts Theo in the nose so hard I can hear something break.

"Fucking asshole!" Theo snaps, grabbing at his bleeding nose. I start charging at Brent, but Kelly grabs my arm, stopping me. Theo looks over at me and holds up his bloodied hand, gesturing that he can handle Brent himself. Theo swipes the blood from his face and flicks it onto Brent's shirt. Then he swipes a finger through the blood underneath Brent's nose, jabbing his knuckle into his face. Brent groans, then winces as

Theo paints the word 'TRAITOR' on the lifebuoy around his neck.

"Brent tossed all the remaining fish into the ocean. Now we have nothing to eat," he says, eyes on Brent.

Xander, from behind me, shouts, "What? How will we survive?"

"We will do what we have done before—figure out another way," responds Theo.

"Kill him, kill him," shouts Karissa so loudly near my right ear that her voice almost bursts my eardrum. Her chanting sets off a chain reaction and the remaining passengers join in, chanting the same words, "kill him, kill him, kill him." I'm shocked that everyone agrees. But he did throw away our only source of food. I stay silent.

Theo bends down, grabs Brent by the ear, and growls through gritted teeth, "Stand up."

Brent doesn't move and Theo yanks his ear harder, damn near ripping it from his head. Brent still doesn't budge, even though Theo is pulling with such force that I can hear cartilage cracking. Right before the ear is torn clean off, Brent stands groaning in pain.

"This is your last chance to say anything," says Theo, his expression cold and emotionless.

"I'll see you all in Hell," says Brent, barely able to get the words out.

Theo steps back and stares at Brent. Then he charges at him and shoves him in the chest. My eyes go wide as I watch Brent's body fall backward over the railing. Before the rope catches him, Brent screams. The rope suddenly catches, and all of Brent's weight hangs on it. Dangling from the chandelier, Brent flails like a fish out of water. He kicks his feet frantically and grabs the rope around his neck. Spit and blood fly out of his mouth with gurgling sounds as he struggles to stay alive. The chanting in the room stops, and everyone stares as Brent slowly ceases moving, his arms falling to his sides as he swings back and forth.

A woman's cry from the back of the room breaks the silence, and I turn to see who it is. It's Megan, a woman who never said much but was always by Brent's side. I watch as her bloodshot eyes lock onto Theo, tears rolling down her rosy cheeks. She lets out a loud scream and charges toward Theo. His eyes widen as she quickly approaches him. Within a split second, just before she reaches him, he steps aside and her momentum takes her over the side of the railing. I watch in horror as she flips over the balcony, screaming as she falls. Her scream is silenced a second later by a loud thud. We all rush to the railing to look down. The woman's body lies face up, her lifeless eyes staring directly at us, and I can't shake the feeling that they are locked on mine. The broken wood railing is pierced through her torso. Blood pours from her mouth.

The sound of blood dripping from the tip of the wood hits the white-tile floor, echoing throughout the atrium like a leaky faucet. I turn to look at Theo to make sure he's okay, and I watch him take a step back from the edge with a blank expression. His face trembles as he looks down at his bloodied hands.

Nobody dares to speak or move. I look around at all the people silently watching my brother and count— one, two, three, four, five, six, seven, eight, nine, ten, eleven. Then I stop, because there is no one else to count. Only eleven passengers are left alive on the ship.

36

THEO

The remaining passengers huddle together on the 6th floor as the frigid temperature drops even lower with each passing second as the sun sets.

"What are we going to do for food?" asks Kelly.

"I'm not sure yet, but we'll figure something out," I say, wiping the blood from my face with my sleeve. All at once, everyone starts to panic at the possibility of starvation.

"I have an idea for you, Kelly," Mary whispers. Kelly and Tyler both look over at her. No one else in the room except Julie and I heard her.

"Okay, but why only for me?" Kelly asks, looking genuinely confused.

"Because we want to help protect the baby as best as we can," Mary answers.

"Well, that doesn't change the fact that we all need food, Mary, so your idea had better be good enough to feed everyone," Julie snarls. I can't help but notice Julie's little power play with Mary is back again.

"Julie, let's hear what she has to say. She's right; we do need to make sure the baby is healthy," I say. I can't help but notice as Julie sucks in a breath and clenches her jaw.

"Thanks, Theo," says Tyler, and I watch him give Julie an agitated look.

"My idea isn't going to please everyone, and it might not even satisfy Kelly," Mary responds.

"Alright, what is it?" Julie asks.

"The baby formula. I know there isn't much, and Kelly should be able to nurse once the baby is born, but we should save most of it in case she can't. If your timing was right and you think you got pregnant in June, which, by the looks of your stomach, I'd say it is, then you will be having the baby sooner rather than later, and he or she will need to eat. Since the formula is powder, maybe we can melt some snow and store it as a backup. The formula will give you enough nutrients to keep you and the baby healthy," says Mary.

"What about the rest of us?" asks Cyrus. Even though he doesn't talk much, knowing he was all buddy-buddy with Brent and Logan makes me not trust him. The way he lifts his brows when he asks irritates me, and I am already on edge after everything that just happened.

"I just said we will have to figure it out; the formula is for Kelly and the baby. Is that clear?" I say.

"Yes, sir. I was just asking," says Cyrus, now with a softer tone.

"Listen, this is a stressful situation for all of us. But I will do my best to help us find a solution to the food problem," I say loud enough for everyone in the room to hear. "With that in mind, if I catch any of you even getting close to the baby formula that Mary has stored away, I will cut off your hands and kill you myself. Do I make myself clear?"

The entire room begins nodding, and I catch Kelly lowering

her head. I get that she may be embarrassed, but I don't care. That baby is my niece or nephew, and is my responsibility too.

In the back of the room, Amy, a younger girl who rarely talks, speaks up. Like everyone else, she wears multiple layers of clothing that have begun to tear and are heavily soiled. "What are we going to do about the heat? It's getting colder, and I can't stop shivering no matter how many layers of clothes I put on."

I study Amy for a moment. She is always so shy and quiet that I never really paid any attention to her, but now that she has my attention, I notice her pale skin appears almost bloodless, and her lips are turning purple. I glance at the others, then back to her, noticing she is by far the smallest and most petite of the remaining people.

"That's a good question. And I agree, it does seem like it's getting colder. I'll figure something out; give me a second to think about what we can do." I say, looking around the room for anything that could help us stay warm.

I look back behind me and only see the destroyed atrium and Brent's body, which has almost stopped swinging. "Let's go back to the medical center and find a room where we can try to contain our body heat," I tell everyone, making my way past them to the stairs.

37

TYLER

We stay close as we climb the stairs. As we walk, we pass the same area where I found Julie just moments earlier. I stop and stare at the wooden board leaning against the hole. The rest of the group continues without noticing I've stopped. Kelly also stops when she realizes I'm no longer beside her. Soon after, Sara and Chris stop as well, like a domino effect.

"Babe, what are you doing?" Kelly asks.

"Has anyone been on the 7th floor since we boarded the ship?" I ask.

"No. Theo said to stay off that floor, it's dangerous," answers Chris.

"There is nothing in there; it's all unfinished. And it's most likely even colder without finished walls. Come on, babe, let's go; I'm cold, and my feet are really starting to hurt," says Kelly, rubbing her belly. I agree that it's most likely freezing in there. Although I can't shake the thought that someone has been in there, I'm too cold to think about it now. And as Kelly said, it's most likely too cold for someone to be going in there.

Once we are inside the medical center, Mary is over at the

closet holding the cans of powdered baby formula. She makes her way back out to the center of the room and meets us. We stay close together, trying to keep our body heat contained.

"Hey, Theo, where is the area that stores all the childcare items?" asks Mary.

"If there is anything, it would be in the daycare on the 4th floor, near the customer service desk." Theo scratches his beard, as if he is thinking about something. "Why do you ask? Isn't that where all the baby formula came from?"

"I guess. But I've never been there. Someone brought it all to me months ago. I'm asking because we have all the formula, but we don't have any baby bottles to use after the baby is born." I look around the room, curious if anyone else is listening in like I am. We are all huddled close together to save heat, so they must at least hear the conversation. I spot Julie—her eyes are locked on Mary as she speaks to Theo. I honestly don't understand why Julie has turned into such a bitch lately. I mean, she has always been a bitch to me, but why be a bitch to Mary? She is the only one helping all of us.

"I'll have someone go check to see if there are any in there," responds Theo.

"I can do it," Mary replies.

"I don't want you to go alone. Take someone with you," says Theo. I watch as he looks up and notices Julie staring at him. "Take Julie," he says. I look back at Julie and catch her rolling her eyes.

"No, I'd rather stay here," responds Julie.

"Come on, Julie, I could use the company," says Mary with a slight grin on her face. I close my eyes and pull Kelly a little closer to me for warmth. I don't care right now about whatever it is those two have going on. I should volunteer to go find the bottle since it is for my baby, but I get the feeling Mary and Julie need to work out whatever is happening.

"Fine, I'll go," Julie finally answers.

"We'll be back. Y'all stay together and try to figure out how we can make more heat to stay warm," says Mary, as she and Julie walk out the door.

38

MARY

The ship is freezing, and my skin is so dry and chapped that it feels like it's cracking. The moonlight filters through the windows, lighting our path through the ship. As we descend the stairs, we pass a broken board on the 7th floor. I don't stop, and if it weren't for Julie staring at it, I don't think I would have even noticed it was there. I look at her and watch as she stares, as if memorizing the placement of the wood.

"Let's talk," I say, getting her attention.

"I don't want to talk, Mary," Julie answers abruptly.

I don't know what is going on with her; I've never done anything to her but try to help. "I don't really care if you want to or not," I say.

"Mary, I said no. Now leave me alone," Julie demands. I stop and stare at her. She pays me no mind and keeps walking away.

"Enough, we need to discuss this," I say sharply. It must get her attention because this time she turns to face me. I'm standing several steps above her now, and she leans against the railing and looks at me, looking almost bored.

"Okay, Mary, what's on your mind?" she asks sarcastically. I make my way down the stairs to meet her. The smug look on her face pisses me off.

"Really? You know exactly what I want to discuss," I say, matching her tone.

"Okay, what about it?"

"What the hell is going on?" I ask. "Why are you being so secretive? And where the hell do you go sometimes? There have been plenty of times when you'll be gone for hours, and no one can find you."

"What difference does it make where I go?" Julie responds, waving a dismissive hand.

"Julie, you can't just brush this off. Between your disappearing and the Theo situation, there is a lot to unpack here. What are you going to do about Theo?"

"I don't know yet; so much has changed. This was not part of the plan," responds Julie, only this time her tone is soft, almost sympathetic.

"No shit!" I say, letting out a deep breath. "All of this is... a mess." I sigh.

"And now there's Kelly and this fucking baby. If Tyler had just kept his dick in his pants, none of this would've happened."

We stand there in silence. I swear Julie is bipolar. First, we are all buddy-buddy, and she told me all kinds of personal things, and then the next minute it's like she wants to slit my throat. I can't keep up with the constant mood swings.

"Leave Kelly and the baby out of this."

"That's easy for you to say, Mary; this is my life, not yours."

"Look around, Julie. This is all of our lives." I gesture around. Her self-centeredness is astonishing sometimes.

"Mary, please," she says, rolling her eyes at me. I feel heat rise to my cheeks in aggravation.

"No. Why don't you tell Theo the truth? He deserves to

know," I say, trying to sound sympathetic, hoping it will change her attitude. I watch her silently as she stares at the floor. After several seconds, she sighs, letting out a deep breath.

"I'm going with my original plan. I can't tell him. It's easier." Then she starts walking away.

"How? Nothing is the same anymore. Just tell him or tell Tyler," I demand.

Julie ignores me and keeps walking. The 4th floor comes into view, and the body of the woman who fell to her death lies at the base of the broken railing. Brent's body hangs from the chandelier, swinging with the ship's rocking, and the rope emits a low, creepy squeak with each tilt. I look away, trying my best not to think about him. I look around and notice... wait, where's the daycare room? The furniture is piled high, blocking the view of the service desk. I look back at Julie again and her breathing is rapid, and beads of sweat are forming on her face even though it's freezing. Damn, she is having a panic attack. I head over to her, reaching over to touch her shoulder. "Hey, Julie, let's head bac–"

She throws her hands up. "I said no! We are sticking to the plan. Now, let's find these fucking baby bottles already."

"Why?" I ask, confused about why she cares so much right now. Julie spins around and stares at me. She doesn't speak. "If you don't tell him the truth, Julie, I will." This instantly gets a reaction.

Julie's expression turns cold as she stares at me. I don't move or say anything, and I patiently wait for her to counter my demand.

"If you so much as speak a word to him, I'll kill you myself. Don't *ever* threaten me again."

I can't do anything but stare at her now. But I'll be damned if I say anything else to set her off.

"Okay," I say, throwing my hands up and stepping back.

"I'm not going to tell him, but I do believe Theo and Tyler deserve to know the truth."

"I've been writing a letter to Theo, and everything's in there. I'm going to give it to him when the time is right." Her tone is no longer angry, as if she flipped a switch and changed her attitude. "Do you have what I asked you to make?"

"Yes, I have it hidden in the medical center. Two, just as you asked," I answer.

"Good, keep them hidden. When the time is right, I'll let you know, and you can give it to both. In the meantime, make one for Kelly, too."

"Are you sure?" I ask, feeling a knot form in the pit of my stomach.

"Yeah, I'm sure," she says as she heads toward the daycare room.

Oh my god, Julie. What are you doing?

39

JULIE

I should have just come to find these stupid fucking bottles on my own. If Mary tells me what to do one more time, I don't think I'll be able to hold back. She knows exactly how to push my buttons and acts like we are the best of friends. I regret ever confiding in her.

"Let's find what we came here for," I grumble, staring at the massive pile of broken furniture that's piled against the wall and blocking the hallway.

Mary doesn't say anything, which is for the best. Instead, she is silently sulking, grabbing broken pieces of furniture and tossing them off to the side. I walk closer to her and grab the leg of a broken chair and start pulling, but it doesn't budge. Mary grabs the other side, and we pull together. After a few tugs, we manage to get the chair unwedged, which causes a domino effect of furniture toppling to the floor. A door past the customer service desk comes into view. It's the childcare center.

Once inside, I'm not convinced we will find what we are looking for. The entire room is in ruins as we navigate through the mess. The dim room is scattered with broken furniture and toys. A loud squeak startles me, and I turn to see what it is. I see Mary lifting her foot off a rubber duck.

"Watch where you are walking."

The room is almost completely dark, and the door is propped open just enough to let in a sliver of moonlight from the atrium. Both Mary and I search through the rubble for any signs of baby bottles, but neither of us has any luck.

"Julie! Guess what I just found!" shouts Mary enthusiastically.

"Let me guess.... a baby bottle?" My sarcasm isn't lost on Mary, because her smile falters before she responds.

"No, even better."

"Then what?" I say, lifting my hands.

"A bag of chips."

Holy shit. I quickly run over, nearly tripping in the mess. When I see the small red bag in Mary's hand, I reach out and snatch it. My mouth is watering as I start to open the bag, but Mary walks over and snatches it back from me.

"What the fuck, Mary? What are you doing?" Mary stands there staring at me.

"We should take these back and share them with everyone," she states.

"Give me the damn bag. What do you think a small bag of chips will do for a group of people?"

"We can give everyone some. They would want some too. This isn't fair for us to secretly be eating while everyone else is also hungry."

"Well, Mary, life isn't fair. If it was, I wouldn't be stuck on a fucking ship in the middle of the ocean, freezing my ass off looking for baby bottles for my stupid-ass brother-in-law's bastard kid, but here we are. So, unless you want to cause a full-blown riot, I'd eat the chips now."

"Okay... I guess that makes sense," says Mary, handing me the bag. I sit down on the floor and rip it open. The cheesy smell wafts into my nose. It takes everything in me not to grab a handful and cram them all in my mouth at once. Instead, I take out one chip and savor the taste. Mary takes a seat next to me and does the same.

When the bag is empty, we both lick our fingers clean. I toss the bag aside with the rest of the mess and stand up. In the distance, I see a plastic bag. I walk over, pick it up, and notice it's filled with kids' sunscreen, unopened pacifiers, and a few other things. At the bottom of the bag, there are two packs of baby bottles still wrapped in cardboard and plastic. I can't help but laugh because I honestly didn't think we would find it.

"Well, I found a bottle," I say, while holding up the pack.

"Good, now let's head back to the medical center, I'm freezing," Mary says.

The medical center is noisy as everyone discusses ways to stay warm. When we walk in, I hear Theo and Sara disagree with an idea someone has that involves a fire inside. They both say that either the smoke will be deadly or it will cause a fire that would destroy the place. Which is a good point.

Mary and I approach the group as they go on and on with ideas. Mary holds up the pack of baby bottles for everyone to see, "Look what Julie found."

Kelly is the first to notice and smiles back. Kelly walks over, hugs Mary and me, and thanks us for going to find it. I then walk over to Theo, smile, and kiss him. He pulls back instantly and stares at me, confused. Seeing his reaction, I'm also confused.

"What? Why did you pull away like that?" I murmur.

He doesn't answer me. Instead, he takes a step closer to me. He stares at me, then at my lips.

"Where did you find food?" he whispers in my ear, making sure no one else can hear. My mind races—how in the world could he know we found food? I've been standing here, and I know Mary didn't tell him.

"What are you talking about?" I say, doing my best to look dumb.

"Julie, I'm not stupid. I can taste it on your lips," he says. Then he lifts his hand and wipes a finger across the corner of my mouth. "You also left proof," he says, showing me the orange residue he wiped off my mouth. "Why are you lying to me?"

It's okay. I can fix this. I know exactly what to say. "I'm sorry, babe. We found a small bag of chips in the childcare room. When I found it, I told Mary everyone was hungry, and we should bring it here to share. But she argued that there wouldn't be enough for everyone and that it could cause a big fight. Then she opened it and suggested we eat it so it wouldn't go to waste. I was starving and couldn't resist, so I agreed, and we ate it. I'm so sorry," I say, my voice trembling as I rest my head on his shoulder. I sniffle a little to make it seem like I'm crying. I look up and see Theo's eyes fixed on Mary, a scowl on his face. Mary notices and smiles at him. He doesn't smile back. He just turns away.

"Thank you for letting me know. You didn't do anything wrong. She shouldn't have put you in that situation. I wonder if she has food hidden anywhere else."

"No, Theo, it's okay. I'm sorry. We're all starving, and it was just sitting there. I don't think she meant anything by it. She didn't want others to get too excited. Please don't say anything to her. I told her I wouldn't tell you," I beg, with a sympathetic look on my face.

"Alright, I'll let this slide. But no one should hide food from others. If it happens again, let me know."

"Okay, I will. I love you," I say, giving him another kiss.

"I love you too."

In the center of the room, Xander jumps up, shouting that he has an idea for heat.

"We have access to a patio out here, but it's too cold to sit around a fire outside," he explains.

"Right, and we can't have one inside because the smoke will kill us," says Theo.

"But, this medical center used to be a spa, right?" Xander asks.

"Yes," responds Mary.

"Okay, good. Most spas have a ton of stones used in their massage rooms. What if we build a fire right outside the medical center on the patio and put the stones in it to heat up, then bring the red-hot stones in for heat and have everyone huddle around them?" Xander says with a huge smile on his face.

"That might actually work, Xander. What a great idea," Sara says. Grateful smiles and hugs spread across the room in excitement.

"Okay, let's find the stones," Theo says.

"They are in the back in a bucket on the ground," shouts Mary, and Xander stands up and rushes over to grab them. After a few moments, he walks out, pulling a heavy bucket filled with shiny black stones.

"What are we going to make a fire out of if there is no more wood in the saunas?" asks Tyler. It's a great question. I look around the room, and there isn't any wood or paper to burn.

"We have all the broken furniture we can burn," says Theo.

"We also have all the books in the library," says Mary, and I can't help but notice that when she says it, she is staring at me. I give her a dirty look.

"We are not burning the books," I snap in reply.

"Let's start by burning the broken furniture. Everyone, walk around the ship, gather as much wood as you can, and bring it back here. Tonight, we'll have heat," Theo shouts eagerly. The room erupts in celebration.

40

THEO

Within minutes, sections of broken furniture start coming through the medical center's door. Behind the help desk in the back room, Tyler, Xander, and I use axes to dismantle the stone wall facade. After removing all the pieces, we carry them to the patio and place them in a circle on the deck, creating a barrier to prevent the fire from spreading or burning through. Once the fire area is constructed with multiple layers of stone, we rush back inside to escape the freezing weather. After our hands warm slightly, we go back outside to carefully place the broken furniture pieces in the fire pit. I look up while we are setting all the pieces and see the rest of the group watching us through the glass doors. It doesn't take all three of us to set the wood in place, and Xander insists he and Tyler can finish it. I try to stay and help, but Tyler insists I go inside. I step back inside and instantly hear a loud voice coming from the back of the room. It's Cyrus.

"How will we heat the stones? Has anyone considered it?" he asks.

Heads turn, scanning the room for an answer. Silence falls when no one responds, and the question hangs in the air.

"We could use the grates from the fire pit," replies Karissa. Before I can share my thoughts, Sara cuts in.

"That's a great idea. Who do you want to take with you to get them?"

I can't help but notice that Sara volunteered Karissa to make the trek instead of doing it herself. I stay quiet, more interested in how the situation will unfold.

"Umm, I guess anyone who wants to go can go," Karissa responds, sounding a bit surprised.

"I'll go with you," says Chris, right beside me. "I'm not sure how to get there, so I'll follow you."

"Okay, that's fine," answers Karissa, gathering more clothes to layer on.

"Thank you, Chris," says Sara, and she gives him a kiss.

As Karissa and Chris leave, I look around the room, and my eyes land on Kelly as she opens a can of baby formula. An audible pop echoes through the room. Everyone turns to see where it came from. Like zombies, we watch completely mesmerized as Kelly scoops the powdered formula into a cup, then adds a little water. No one moves or makes a sound during the entire process. It's been almost three days since anyone has had anything to eat or drink, and hunger is setting in with a vengeance.

"Can you not do that in front of us?" shouts Amy from the back. Kelly's face instantly turns red and she puts her head down. I'm also a little shocked because Amy rarely speaks.

A cold breeze from the open door to the patio sweeps in as Tyler and Xander step inside, sending a chill down my spine.

"I'm sorry," says Kelly softly, and I hear the fear in her voice.

"The least you could do is take it somewhere else. If I had known that spreading my legs would have gotten me priority at mealtime, I'd have fucked my way up the food chain too," Amy says, directly to Kelly.

"Hey now, what's going on here?" I say.

"Nothing, I'm fine," responds Kelly.

"Who the hell do you think you are, talking to her like that?" Tyler shouts at Amy.

"Tyler, it's okay. I'm sorry, everyone. We have about four cans of baby formula I can share, I really don't mind," Kelly says loudly. I can still hear her voice trembling with fear.

"I said no! The formula is to keep you and the baby healthy. I am not going to say it again," I growl.

"I can solve that problem," Amy shouts, and I watch her look toward the wall where the stone wall has been deconstructed.

"And how exactly do you plan to do that?" Sara asks from across the room in a tone I haven't heard from her before. It's a little frightening. Amy doesn't answer and slowly walks away.

"Everyone needs to calm down. We have to stay together to keep warm if we are going to get through the next couple of days. It's the end of February, and the temperatures should start to rise soon. When they do, hopefully the fish will begin biting again," I say.

"Shut the fuck up with your bullshit," shouts Amy.

The room erupts in conversation and arguments. While everyone else is going on, I ignore Amy's comment, knowing she's only looking for a fight. As Amy walks away, I turn and notice Mary standing in a corner alone, her eyes fixed on something. When I follow her line of sight, I watch Julie quietly leave the room. I wonder if she knows Mary is watching her too. I look back at Mary and realize she doesn't know that I'm watching either of them. Within seconds, Mary slips out right behind Julie.

The remaining passengers in the room are too preoccupied with their own conversations to notice them leaving, and I'm tempted to follow, but I have a gut feeling Amy isn't finished causing trouble. I watch as she reaches the wall behind the desk, then she stands there silently, unmoving. I'm the only one who is watching her.

Suddenly, she starts running toward Kelly, an axe raised above her head. But before I get the chance to warn Kelly, she hears Amy, because she turns around lightning-fast and stands still, as if paralyzed with fear. Amy, without hesitation, throws the axe as hard as she can.

Tyler, standing beside Kelly, pulls her out of the way. They both plummet to the ground. Tyler's arms wrapped tightly around her in an attempt to break her fall. Amy stops running as Kelly looks up at the axe that is now embedded into the wall above her head. Kelly doesn't speak or move. Her eyes start rapidly blinking as she looks around, and then she looks back at Amy and back to the axe. Her brows furrow, and I see her jaw clench so tight she could break a molar. The look in her eye says it all—Sophie's back. She hefts herself up and stands as quickly as she can. I watch Tyler reach out, trying to stop her, but he's no match for Sophie when she comes out. Sophie yanks her hand away, grabs the axe from the wall, and sprints toward Amy.

The gap between her and Amy closes within seconds; she's moving so fast that no one has a chance to stop her. Amy's eyes widen as she watches Sophie lift the axe high and bring it down. The cutting edge of the steel axe slams directly between her eyes, embedding deep within her skull. Her nose splits in half and blood pours down her face. Her arms start to flail as her legs give out, and she falls. Sophie turns to face us, pulling the axe out of Amy's head in the process. She lifts her hands and wipes the splattered blood off her face, then spits on the ground.

"If anyone fucks with my baby or me again, I'll kill you. I

don't care who you are." She looks around the room as we all stare at her. "Someone call my husband. I'm ready to go home."

"It's okay, Sophie. I'll call him for you," Sara says, slowly walking up to Sophie. "While we wait, I have someone I want you to meet. His name is Tyler. Your husband told me they work together and he sent him to take care of you until he gets here to pick you up, okay?" Then she turns to Tyler, muttering, "Just follow my lead."

I rub my eyebrow. I don't know if I will ever get used to this craziness.

"Yeah, he is a great guy; he should be here any moment," says Tyler, catching on quickly. Tyler stands and slowly approaches her. He stops before reaching her, looks Sophie in the eyes, then shifts his attention to her hand holding the axe. Sara slowly reaches out and takes the bloody axe from Sophie's hand.

"Hi, Tyler. I'm Sophie," she says in a high-pitched, giddy voice. I can't help but think of a crazy lady in a slasher movie, with the way she's talking, and the blood splattered across her face.

"It's nice to meet you, Sophie. I've heard a lot about you. Let's find somewhere to sit. How does that sound?" Tyler keeps his tone light and calm.

"That sounds nice," Sophie responds, and then they make their way to the back of the room and sit on the floor, where blankets are piled up. I watch them for a moment as they sit next to each other. I'd be terrified in this moment, but Tyler looks more curious than anything.

"Why don't we try to get the fire started and let things settle down here?" says Xander, pulling my attention away.

"That sounds like a good plan. Give me a moment," I say, walking over and grabbing a sheet from the back of the room and draping it over the body. Instantly, blood seeps through, staining the dirty white cotton crimson. At this point, I'm numb to the sight of blood and dead bodies. I grab the legs after I wrap Amy's body in the sheet and drag her out of the room and up to the top deck.

41

MARY

Quietly, I follow Julie down the cold, dark hallway, watching her every move. Carefully, I avoid making any noise, maintaining a safe distance. First, I watch as Julie enters the library on the 6th floor. I'm doing my best to stay unnoticed, so I wait until the door closes, then peek through the glass. I stare into the library and watch as Julie looks around nervously. Then she removes a book from the shelf, reaches back in, and pulls out another from deeper inside the shelf. She opens it and takes out folded papers.

"The letter," I say aloud, then quickly cover my mouth, realizing I may have just blown my cover. I peek from the side of the window and see Julie facing it. I drop back down and pray she didn't catch me. I peek through the window one more time and see Julie shoving the letter back into the book and placing it back on the shelf. Carefully, she arranges the pile of books in front of the one containing her letter. I start to panic as Julie rushes out of the library. I'm not sure where to go. The ship sways in the water, causing a slight tilt, and the sound of an eerie creak fills my ears as Brent's body swings from the chandelier. I hear the door to the library begin to open and make a run for it, sprinting toward the stairs.

After reaching the stairwell, I hide behind the wall as the squeaking library door slams shut. A moment later, I hear footsteps—receding rather than approaching. Feeling a bit relieved, I exhale a breath I didn't realize I was holding. Carefully, I peek around the corner and watch Julie glance around, then take off in the opposite direction. I consider whether to follow her or search for the letter. No, I can come back for that. Right now, I need to see where Julie is going. I head in Julie's direction and hear her descending the stairs. I follow her all the way down to the 1st floor, which is shocking since there isn't anything down here, and it's completely dark. Julie pulls out a small flashlight and turns it on, shining it all over.

I reach into my pocket to make sure I still have the flashlight I found a few months back. A moment later, Julie and her light disappear behind a steel door labeled '*Crew Only.*' The room goes dark, and I click on my flashlight. I have no idea why Julie would need to be down here, and I plan to figure it out. Fifteen minutes pass, and just as I'm about to give up, the steel door opens, drawing my attention. I click my flashlight off instantly as Julie runs out toward the stairwell. I stay hidden behind a nook in the wall. Once she is gone, I rush to the steel door, opening it quickly and quietly shutting it behind me. There's a foul odor lingering in the air. The room is stuffy, and right away I feel claustrophobic. My mouth drops when I shine my flashlight around the room. Trash is scattered across the floor, including empty cans of beans and boxes of dried food.

As I press on, I notice empty water bottles scattered around. *No wonder she's always hiding and never complains about food; she's been secretly eating all this time.*

While walking, I stumble upon a pile of bones and let out a scream, then quickly cover my mouth, hoping no one heard me.

I step closer to get a better view and figure out the decayed corpse is Captain Whitaker. His nametag is on his uniform. I don't want to be alone down here any longer, so I turn around and head back toward the door. As I'm leaving, a bright yellow object catches my eye—a bag of some sort. Approaching it uncertainly, I notice there are two bags, each with red letters printed on the side. I can't quite make out what they say.

The closer I get, the more visible the letters become. In large red print, I make out the letters 'SURVI— The screech of the door opening has me dropping to the ground. I try my best to turn off the flashlight, but my nerves have me fumbling with it. I quickly jam it into my thigh to hide the light. The door slams, and I finally manage to flip the switch, turning the flashlight off. I wince at the audible click from it.

"Hello, is anyone there?" someone whispers. I realize it's Julie.

I'm unable to move out of fear of being caught. I hold my breath and remain perfectly still. As soon as Julie leaves, I exhale slowly and rush to the door. Feeling trapped, I fumble with the flashlight, unable to switch it back on. My heart is pounding, and I feel like I'm going to pass out. Eventually, the flashlight comes on, and the door is only a few feet away. I run and swing it open, no longer worried about getting caught. I've got to get out of this room. As soon as I'm on the other side of the door, I jump as I hear the door slam behind me. I can't help the chill that runs through me. What the hell is going on here? No longer wanting to be anywhere near the 1st floor, I sprint up the stairs, out of breath, not stopping until I'm at the entrance of the medical center.

I walk into the medical center and immediately spot Julie off to the side, talking to Karissa and Sara. Kelly or Sophie, whoever she is now, is beside them. She is talking so loudly that the entire room can hear her say that her husband will be there any minute to get her. Julie looks up, and our eyes meet briefly. She smiles, then turns back to the conversation, as if she wasn't just sneaking around. I begin to approach her, intending to ask why she was sneaking around, but my attention is pulled by a loud scream from outside.

Xander, Theo, and Tyler are jumping up and down excitedly as the fire roars to life. Carefully, they line the stones across the grates. Once all the stones are placed on the grate, they set it over the fire to heat. I watch Theo after they walk inside. He carefully removes his face mask, and when he looks up, he catches me staring at him. I'm surprised when he scowls at me for the second time today. What in the world have I done to make him mad? He turns away to walk toward Julie, and I decide to talk to Julie later when she is alone.

I take a seat near the window to feel the sun shining in. About twenty minutes pass, then I see Xander stand and make his way outside to collect the hot stones. Theo and Tyler follow, and together they carry the metal grate filled with hot stones indoors and place it over a metal bin. Everyone quickly crowds around, and Theo grabs a heavy blanket and tosses it over us to help trap the heat.

"I forgot how much I missed the heat. I can't wait to be back on land and bask in the sun for hours," says Karissa.

"You and me both, sis. I never want to see the ocean or a ship again," responds Xander. The group erupts in laughter, and for that moment, I smile along with everyone. For that brief moment, everything feels normal for the first time in months.

"What is everyone laughing at? Why on earth would you not want to see the ocean or a ship?" asks Sophie, oblivious to the entire situation. I can't help but wish my mind were unaware of what's happening. I push that thought away, knowing it isn't possible.

"Don't worry about it, Sophie. He is just being silly," replies Sara.

"I don't get the joke, and why is it so cold? Could someone turn on the heat?" says Sophie.

We ignore her, hoping Kelly will come back soon, and savor the small warmth we share. The heat lasts only so long before we go back to sitting around the room, trying to keep ourselves warm while we try to fall asleep for the night. I wish I could just go to bed and wake up anywhere but here.

Several hours pass, and the morning sunlight streams through the medical center's windows. A cold, wet sensation wakes me. As I open my eyes, I see the remaining passengers all lying motionless on the floor. Each one shivering—the heat is gone. I look outside and notice the fire has gone out. The sound of light snoring surrounds me while everyone sleeps. Curious if anyone is missing, I count each mound on the floor. I'm interested to see if Julie has disappeared again. Tyler, Kelly-Sophie, Sara, Chris, Theo, Julie, Karissa, Xander, Cyrus, and I. Ten people; no one is missing. Sitting alone, I think back to yesterday and the 1st floor, where I nearly had a panic attack. I remember the trash of opened dry food scattered across the floor. I glance over at Julie, who's still sleeping, and the letter she was holding in the library comes to mind. Now is the only time I will have to find the papers and see what she wrote. Without wasting any more time, I quickly head for the library.

The hallway is cold, and frost covers the windows. I shiver as I exhale, and a cloud of my hot breath billows back into my face. An eerie squeak echoes throughout the ship with each tilt on the rough waters, making me just a bit more on edge. Quietly, I descend the stairs. Knowing I have only a few minutes before everyone begins to wake up, I quicken my pace. When I reach the library entrance, I open the door, and a loud creak echoes throughout the atrium. The musty smell of mildew nearly takes my breath away. Without stopping, I walk directly to the bookshelf where I watched Julie hide the book. Not exactly sure where she hid it, I pull several out at once, searching for the hidden one. Frustration grows as I pull all the books off, tossing them onto the floor. After several minutes, I get frustrated looking around.

Having no luck in finding it, I decide to go back to the 1st floor and get a better look at what I saw down there. Standing up, I exit the library to head toward the 1st floor when a sudden thump startles me and I stop suddenly. The sound came from above me. I wait for a moment, hearing nothing but the quiet creaks and groans of the ship. But something draws me up the stairs instead of back down. Needing to figure out what it was, I look around the 7th floor, and there is nothing.

"Hello?" I whisper. No response. "Hello? Is anyone out here?"

There is still no answer. As I look around the 7th floor, I notice a piece of wood that had once been propped against a hole now lies on the floor. It's the hole that Theo created months ago. Oh, that must have been what made the noise, the board falling over. I feel a bit less freaked out now, but not by much. I inhale and exhale, taking a seat to try to relax and slow my racing heart. Resting against the wall, I close my eyes and lean my head back. A slight noise behind me startles me once more. I turn and look at the hole. Something is moving inside. I need to know what's in there, so I push my fear away and climb inside.

42

TYLER

The medical center floor is freezing against my skin. I squint against the blinding sunlight and roll over. Kelly is sleeping quietly beside me. I look down at her belly and smile at the thought of becoming a father. Then, just as quickly, my smile vanishes as I look around and remember we're still trapped on a ship. I reach out to rub Kelly's stomach, then hesitate—unsure what Sophie will do if it's her who wakes up and feels me touching her. I look at her belly again and reach anyway; a small kick sends me flinching, and I pull my hand away. A wave of emotion floods through me, and a tear rolls down my face. Kelly lifts her hand and wipes the tear from my cheek. I look her in the eye, and she cups my face.

"Good morning, babe," she says softly.

"Kelly?" I whisper, unable to hide my excitement that it's not Sophie. I can't help but notice the look of confusion on Kelly's face at the question. She looks over to see the blood on the floor where Amy's body was.

"What is that from?" she asks, pointing to the blood smeared across the floor.

"It's nothing, try not to worry about it."

She gasps; tears start to stream down her cheeks.

"Sophie did that, didn't she?" I turn away, not wanting to answer, because I'm not sure what exactly to say.

"I need help," she whispers. At those words, my heart melts, and I feel so bad for her, knowing there isn't anything I can do.

"Babe. It's not your fault. You didn't do this. You have no control over when Sophie comes out. I'm not saying it's okay, but you can't beat yourself up. If it makes you feel better, I've spent a lot of time with Sophie, and I think she's warmed up to me. I'm no longer a threat to her or *her* baby."

"*Her* baby?" asks Kelly, her eyebrows raised.

"Yeah, funny thing about that. Sophie truly thinks the baby is hers and her husband's. She also believes her husband is about to pick her up as soon as he gets off work," I say softly. Kelly frowns and looks as if she is about to start sobbing.

"I'm sorry. After all these years, why is it happening all the time now? I feel like I'm going insane," she says, her voice trembling.

"Because we are in an extremely high-stress, life-threatening situation, exactly what your doctors claimed would will bring Sophie out."

"Tyler. What if I hurt you or our baby?"

"You can't think like that. I really don't think you will."

"I hope not," she says softly.

I hear shuffling from across the room, and I look over to see people waking up.

"Good morning, everyone," says Theo hoarsely as he stretches.

"Good morning," Julie responds. Then everyone else responds the same way.

"Fire is out; we need to get that started again," says Xander as he stands. "I'll do it now. Where is the lighter we found?"

"Should be over next to the pile of broken wood and furniture," answers Theo.

Xander heads out to the patio. As soon as the door opens, a blast of icy snow and freezing air rushes in. The cold shock instantly perks me up and, evidently, everyone else too.

"Does anyone know where Mary is?" asks Julie. I look around the room and I don't see her anywhere.

"No," I say, and then multiple people around the room say the same.

"Where do you think she is?" asks Theo, and I notice he is directing his question to Julie directly. He asks with a look that says he is asking something only Julie would know.

"Honestly, I have no idea," replies Julie.

"Maybe she just went for a walk. Everyone is under a lot of stress," says Sara.

"Yeah, you're right," responds Theo.

"That sounds like a great idea. I am going to stretch my legs too. I'll look for her," responds Julie, with a bit too much excitement in her voice this early in the morning.

The door opens again as Xander walks in, sending another blast of snow and freezing air.

"I got the fire going again. Let me get those stones, please," says Xander, rubbing the palms of his hands together to warm them.

Everyone moves around, making way for Xander to carry the grate of stones back outside. I have a strange feeling that something is wrong but I'm not sure what it is. I glance over and watch as Julie exits the medical center. I stand up to head after her, but Kelly grabs my arm, pulling me back.

"Please stay here with me," she asks. I don't want to upset her, knowing she has had a rough night, so I sit back down and let my thoughts run wild while she and Sara talk about their past.

"Everything okay, babe?"

"Yeah, I'm fine. What are you and Sara talking about?" I ask, even though all I can think about is where Julie is going.

43

MARY

I crawl through the hole leading into the unfinished 7th floor. I can hardly see because there are only a few uncovered windows, and most of the natural light is blocked. I pull out my flashlight and turn it on. I'm shocked to see the unfinished floor stretches the length of the ship. I look up and see metal scaffolding and hanging lights everywhere. Moving cautiously, I shiver in the freezing air and take in my surroundings.

The hallway is long, lined with walls and doorways that look like they're meant for future guest rooms. Some have sliding glass doors leading to balconies, most covered with dark plastic tarps, while others have solid walls instead of windows. All the rooms are missing their ceilings. The rooms are empty, their floors covered in a blanket of dust and construction material. As I turn around, about to head back to the hole, my flashlight reflects off something in the distance. I walk over to where my light is bouncing off the object like a disco ball, casting light everywhere. The closer I get, the more I can make out that it's a water bottle propped on the framing of what looks like it's going to be a door. My first thought is that it's from the construction crew that was building the ship, because it looks old and there's trash everywhere, along with other bottles that are empty. When I shine my flashlight into the room, I gasp and accidentally drop it. The flashlight rolls as the ship tilts, causing the light to move across the room and scatter reflections everywhere. The floor is covered with full, unopened bottles of water. In the back of the room, there's a small area with two large cans of beans, a manual can opener, and about five boxes of saltine crackers. On the opposite side, at least twenty empty cans of beans is scattered all over.

A noise in the distance pulls my attention from the opened cans. Quickly, I grab my flashlight and rush back toward the hole I crawled through. As I climb out, the broken wood catches and rips my shirt. Once outside, I run up the stairwell, and as I reach the top, I turn around to see if someone is following me. When I turn back, I run directly into Julie, who is coming down the stairs. I fall backward, tumbling down the stairs. After a moment, I move to stand and a sudden jolt of searing pain courses through my nerves. I look down and nearly pass out when I see that my arm is twisted unnaturally. My arm is broken. Suddenly, I'm dizzy and grab onto the wall with my uninjured hand. I look up to see Julie running down the stairs straight toward me.

"Oh my god, Mary, are you okay?" she asks.

"Get away from me, Julie," I hiss.

"What the hell is your problem? I'm trying to help you."

"My arm is broken." I wince as pain shoots down my side.

"What can I do to help?" asks Julie.

"For starters, you can tell everyone about the food you have hidden on the 7th floor."

"What are you talking about?"

"I just came out of the 7th floor. There is an entire room filled with bottles of water and food, and empty cans are scattered everywhere." I pause—the pain in my arm is excruciating.

"You've been feasting while everyone else starves. I knew you were a psychotic bitch, but this... Actually, it makes perfect sense now that I think about it. Just like keeping the damn chips to yourself."

"Mary, stop. That's not mine, I promise," Julie pleads.

"You're lying."

"I don't know what you're talking about," says Julie defensively. *This bitch has lost her mind, like, I can't believe she is lying.*

"How about you explain the wrappers I found on the 1st floor in the crew area?"

"Was that you who was following me?" Julie asks, confirming my accusations.

"Yes, what were you doing down there?" I ask, now in a calmer tone, still wincing in pain.

"Come on, we need to go take care of your arm," she says, as if she actually cares.

"I'm not going anywhere until you tell me what's going on. Why are you keeping things from me?" I bite back a groan. My arm fucking hurts, but I need to know what the hell this bitch is up to.

"Look, I found some dry food in a hidden closet in the kitchen months ago. It's not enough for everyone; it was barely enough for me. I've been going down to get a nibble here and there to keep my energy up."

"Why didn't you tell me? Are we still doing what we talked about, or did you change your mind?" I ask. When Julie stays silent, I continue, "What is that plan going to do now? There isn't a reason anymore, Julie. So much has changed. You won't even—."

"Hey, are y'all okay down there? What's going on?" Xander's voice drifts down to us.

"We're fine. We are coming up now." Julie shouts, then

turns to me. "Mary, please, I promise someone else is hoarding all that food, not me."

I don't believe that at all. "Okay, just tell me what the plan is now," I say through gritted teeth. I take a deep breath, trying to breathe through the red-hot pain in my arm.

"Okay, I will. Why don't we keep what you found between us for now, and then we can see how much is in there later," says Julie. I'm in so much pain at this point that I don't even care anymore.

44

THEO

“What happened?” I gasp the moment Julie and Mary walk in, my eyes locking onto Mary’s arm.

“I’m fine. I fell down the stairs, but I’m okay—it’s just broken. Someone, please grab my bag; there’s a splint and some bandages inside,” Mary says as she walks over to a table. Karissa quickly runs to the closet where all of Mary’s supplies are stored, grabs the bag, and then returns.

“Here, what do we do?” asks Karissa as she sets down the medical bag.

“First, we need to put the bone back in its proper position,” says Mary, and I wince at the thought.

“How do we do that?” asks Karissa.

“Oh, I know,” shouts Xander excitedly. “You just pull hard. I’ve seen it in movies.”

“Yeah, kind of,” Mary answers as she takes a deep breath. “Get me something to bite on to help manage the pain. On the count of three, I need you to pull.”

“Umm, okay, I think I can do this,” says Karissa, sounding like she really isn’t up to the challenge.

"I'll do it," says Xander. I watch Karissa relax as soon as Xander volunteers.

"On the count of three, I'm going to pull, one, two–" Without saying three, he pulls, and the bone snaps back into place.

"I found food," Mary blurts out the second Xander sets the bone. Shock ripples across the room, and I'm not sure whether their expressions are from what Mary just announced or from what's just taken place.

"What did you just say?" I ask.

"She said she found food," replied Sara.

"Is that so?" I say sarcastically, then turn my attention to Julie, wondering whether she is thinking the same thing I am—that Mary is hoarding food. I look up, and the rest of the group is staring at me, surprised.

"What's your problem, Theo?" asks Mary.

"I said how convenient becau—"

"He meant great news," Julie says, cutting me off. "It is great news. She found it on the 7th floor; someone has been hiding it."

"What were you doing up there, Mary?" I ask.

"The better question is how much food?" shouts Tyler from the back of the room, and then everyone else begins to murmur about it.

"Well, I found a couple of cans of beans and some saltine crackers. But someone has been keeping it all to themselves because there were a ton of opened cans that looked fresh," says Mary.

"Maybe it was someone from Logan's side when they had all the food," says Cyrus.

"No, that was months ago, and one of these cans looked like it had been opened recently," says Mary, staring directly at Julie.

"Where did you say you found it?" asks Tyler.

"I noticed the board was knocked over on the 7th floor, and I thought I heard something, so I went inside."

I don't say anything. I shift my attention to Julie, and when she sees me looking at her, she raises her eyebrows in suspicion and then looks away.

The room goes silent as Xander and Karissa carefully wrap the splint around Mary's fractured arm. I turn my attention back to Mary and study her for a moment, contemplating what to say.

"Why are you looking at me like that?" Mary asks me directly. I don't acknowledge her question as I run through all the scenarios in my head. I think about what Julie told me about what Mary did in the childcare room, and how Mary's been able to walk around freely, with no one giving her a second look because she is everyone's nurse. It occurs to me that I have no idea what she does when no one is around. Hell, come to think of it, I don't know what anyone does when I'm not watching. Suddenly, I'm feeling overwhelmed.

I shake my head. "It's just frustrating to think that one of us is hiding food from the others; we all want to survive this. And if I find out who did it—so help me God, I will throw whoever it is into the ocean myself."

No one moves; they all stare at me as I scrutinize each person to see who is acting weird. But no one seems any more suspicious than usual. "Tyler and Xander, come with me. We're going to find the food Mary found."

"I can help," says Chris.

"No. Stay here. We are fine. I don't need anyone else getting hurt," I respond.

The three of us crawl through the hole and look around. Xander and Tyler have never been here before, and they are both shocked by the sight.

"Why was this ship able to sail like this?" asked Tyler. From the look on Xander's face, I'd bet he's thinking the same thing.

"Let's just focus on what we're here for. None of that matters anymore," I say, not wanting to get into those details.

"Agreed," says Xander.

"Over there, I see a bottle sticking out of the doorway," Tyler shouts, pointing down the hall. We walk into the room, and I'm speechless as I take it all in.

"Wow, she was right—someone has definitely been down here recently eating," says Xander, kicking the opened cans around.

"Let's gather everything we can carry first, then come back for the rest," I say.

The entire process takes about an hour to transport all the food and water from the 7th floor to the medical center. The most difficult part is carrying all the water—twenty-eight bottles.

"This won't last much longer. The cans split among all of us will probably last only two days. The crackers will hold out longer, but not long enough if we don't start catching fish again," I say.

"Right, but we've tried repeatedly to catch fish. And it's too cold to stay out for long," says Xander, and Tyler nods in agreement.

"I know, but maybe these will last until the temperature starts to rise and more fish will appear," I say, doing my best to be optimistic.

"Yeah, I hope so," Tyler says, his tone clearly expressing disbelief that it will happen anytime soon. Sadly, I also don't think it's enough, but I have to keep being positive.

When we walk into the medical center with arms full, the

group erupts in excitement, and everyone rushes up to see what we found. I tell everyone to step back and explain that we will devise a plan to divide the portions fairly.

"There isn't much here, so we need to make this last as long as possible," I say.

"It's been a long time since we last ate. Can we have some now?" asks Cyrus. I agree. It's been a while, and we are all starving—literally.

"We can even heat the can of beans over the fire," replies Karissa, and the group grows rowdy, jumping with excitement at the thought of a hot meal.

"Yeah, we can." I take a deep breath and look around at everyone. "But we need to ration this like we do everything else. Just a bite or two per person and a few crackers each. That's it."

The scent of beans boiling over the fire has everyone on edge. No one minds that the temperature is below freezing; they all gather around the fire, eyes fixed on the can, each holding a spoon. Once the beans are warm enough, I use a cloth to carry the can inside. Each person takes a turn dipping their spoon into the can for a bite of beans. Chris is the last to get a spoonful. When he walks up, he trips and falls, knocking the can of beans to the ground. The remaining beans spill across the cold floor, triggering a wave of panic. The group stands in silence for a moment.

"I'm sorry," Chris says, breaking the silence.

"It's okay. We will clean it up and put them back in the can," I say. Suddenly, Cyrus drops to the ground and begins licking up the beans. Then, as if on cue, everyone else drops to the floor and starts eating the beans too.

"Everyone, stop. This is all the food we have," I scream, tempted to drop down and eat the beans myself.

But they all ignore me, their hunger clearly louder than my

pleas. After a few moments, the floor—once stained with dried blood, grime, and beans—is cleaned so thoroughly it looks brand new. Each one stands and wipes their mouth, looking satisfied. I clench my jaw, feeling my jaw muscles flex.

"Is this how we're going to behave? Now what? We only have one can of beans left and a few boxes of saltines. That's all we've got, people!" I shout in frustration.

"I don't know what happened—I just... dropped to the floor and started eating. I couldn't stop myself," says Cyrus, his head down.

"You knew this was all we had, Cyrus," I say, and the room falls silent.

"I said I was sorry," cries Cyrus.

"Sorry isn't going to save us now, is it, Cyrus!"

"Theo, please. He didn't mean anything by it. Don't hurt him," says Tyler.

"I'm not going to hurt him, but as punishment, he isn't getting any beans from the next can. We're also not opening it until tomorrow," I say as I grab the unopened can of beans and walk away.

"I can take the can," says Mary, staring at the can in my hand.

"The hell you are," I answer.

"I can lock it up with the formula," says Mary, frowning at me.

"No, it's okay. I'll hold on to these." The fuck I'm gonna let her, or anyone else, get their hands on this food.

45

TYLER

The fire burns brightly outside on the patio, casting an orange glow against the dark night sky and illuminating the entire room. I sit back and count the remaining passengers: Theo, Julie, Kelly, Mary, Sara, Chris, Karissa, Xander, Cyrus. We're all huddled close together to keep the heat in. No one speaks or moves, and gradually, each person falls asleep—that's the only escape from the harsh reality we live in, and it's the only way to make the days pass faster. I stay awake because it's my turn to keep the fire burning overnight. After the other night, when we all woke up freezing, Theo and Xander volunteered to take turns staying awake to keep it lit and rotate the hot stones. Lucky me—tonight is my turn. I do my best to stay awake, but this heat is making it hard because I am so sleepy. I step outside to let the sting of the cold air wake me up. I walk over and stroke the fire, then make my way back inside. When I step back inside, I wrap a blanket around me, and my body relaxes. After a few minutes, the shock of the cold wears off, and I start to doze off.

I'm blinded by sunlight as I open my eyes. My body is shivering, and I pull the sheet over my eyes and try my best to snuggle up. I jump up, realizing I'm cold because the fire is out—the fire I was supposed to keep burning.

"Shit, Theo, wake up," I say, shaking his shoulder. Theo winces, and his eyes shoot open, staring at me.

"What? Stop shaking me. It hurts."

"Oh, sorry," I whisper.

"What do you want?" asks Theo.

"The fire is out. I accidentally fell asleep," I say, and Theo quickly sits up. He looks around, and I'm not sure what he's looking for. Frantically, Theo's hands move everywhere, and he starts yanking off blankets and sheets from each person, searching for whatever it is he has lost.

"What's going on?" asks Sara in a hushed tone, stretching and waking up.

"Yeah, babe, what's happening? Are you okay?" Julie asks.

"No, everything is not fine," Theo growls as he continues removing blankets from each person. "It was right here next to me. Where is it?"

"Where is what?" I ask.

"The food! It's gone; the beans are gone, and the boxes of saltines are gone. They were right beside me," he snaps, his voice growing louder and his face turning a deeper shade of red with each breath. I step back, preparing for him to explode any second now. I've seen him when he gets this upset before, and the aftermath isn't pretty.

"They have to be here somewhere; we are all here. No one is missing," Julie says.

"Where is Mary?" demands Theo.

"I'm right here." answers Mary, who is standing directly behind Theo.

"Where were you just now?" asks Theo again, accusingly. What does he know that I don't?

"I've been right here the *entire* time," Mary says, her eyebrows furrowed and clearly annoyed by the accusation.

"Someone has taken the food. Everyone, get up now!" Theo demands so loudly that saliva sprays from his mouth as he speaks. We all jump up and do exactly what he says, too afraid to move or say anything. Theo begins throwing whatever he can get his hands on, searching for the food. I look around and see there's no sign of it. In the corner, all the water bottles are lined up exactly as they were the night before. "Tyler, this is your fault. You were supposed to stay awake and keep the fire going," says Theo, his eyes filled with an anger I've never seen directed at me before. I've never feared my brother before, but now there's something about him that makes me feel uneasy.

"I'm sorry, I didn't mean to fall—"

"Theo, don't talk to him like that. I get that you're upset but that doesn't give you the right to take it out on him. If you were so worried about the fire going out or someone taking the food, you should have stayed awake yourself or done what Mary suggested and locked the food up," says Kelly, defending me. I feel small that she has to defend me, but I can't hide that I love seeing her stand up for me.

"Shut up, Kelly, and sit down," shouts Theo right back at her.

"No," I say, jabbing my finger in his face. "Don't talk to her like that. I don't speak to your wife that way, so don't talk to my girlfriend that way. Do you understand?" My reaction must have an impact because Theo steps back, stares at me silently, then looks around the room.

"You're right. I'm sorry, Kelly. That was our only food and now we have nothing left. Someone has taken it all."

The room remains silent as everyone glances around.

"We have gone too long without food. We'll die soon if we don't find some. I don't care who took it at this point, just return it so no one else dies," he says, and I watch his shoulders droop as a look of defeat washes over him.

"If no one comes forward, what will we do?" Karissa asks softly.

"Well, I'm not sure," replies Theo.

"Starve," whispers Julie, just loud enough for everyone to hear.

I close my eyes and take a breath, knowing she is right. I do my best to keep my emotions in check. I don't want to think about dying.

"I have the baby formula; we can share it. I know it's supposed to be just for me, but I won't sit here and feed myself while everyone else is starving. That's not right, and I'm not that kind of person," says Kelly. I smile, watching her. She's so caring and so beautiful. It's just my luck that I finally meet someone I truly love, and it's all going to be taken away from me.

"Yeah, but Sophie is," mutters Chris. We all gasp, shocked that Chris said that.

"I'm sorry, I didn't mean any harm; I just meant that Sophie wouldn't share," replies Chris. It's sad, but Chris is right. If Sophie were here with us, she would never offer to share.

"It's okay, Chris. I get it, and you're right. We need to make sure she doesn't come out again," says Kelly.

"Okay," I answer before anyone has a chance to argue. "But let's only use one can because the baby will need it."

46

THEO

Two weeks have passed, and there's been no food besides baby formula. The last of it has been used, except for a single can reserved for the baby expected any day. Each day, people walk the halls to stay warm and stretch, speaking little and keeping their distance. No one has confessed or returned the stolen food. Tensions rise as everyone weakens with each passing day. According to a marked calendar, if it's accurate, it's March 3rd, and temperatures are beginning to rise. I keep my thoughts to myself, but I'm optimistic the fish will return. Every day, two people go out onto the icy, snow-covered deck to cast a net, hoping for fish, but only ice and trash have been caught. Fear and anger are mounting, and arguments have become common as starvation sets in. I've noticed Mary and Julie keep their distance, speaking only when necessary. I also keep my distance and stay quiet. I'm convinced that everyone is against me and I feel like I'm the only one who truly believes we will be saved. Xander and Karissa are always together, which makes sense. Sara, Kelly, Tyler, Chris, and Cyrus never leave the medical center. With the baby expected soon, Mary has the bag of medical supplies ready. Only she and Kelly know where the single can of baby formula is kept– a safeguard so that no one takes it. With the hunger we are all feeling, people are no longer themselves.

"It's going on day three without any food. We need to open that last can of baby formula, or we're going to die," says Cyrus, and it sounds more like a demand than a statement. Each of us has lost a lot of weight, some more than others. The only person who's eaten more than anyone else is Kelly, which is as it should be.

"No," Tyler responds with a sigh. "That's for the baby; please stop asking."

"Well, if we are all dead, who will be here for the baby?" asks Cyrus. And I can't help but consider his question. He is right. But if Kelly can't nurse, then that is all the food my niece or nephew will have.

"Ask him again, and I'll kill you myself," Tyler snaps. "I've told you before and I'm telling you now, that formula is for the baby." I watch Tyler clench his jaw and take a deep breath. "How about you make yourself useful, go and help Xander with the fire, and leave us alone," demands Tyler.

I'm shocked seeing my brother act like this. At the same time, I'm also proud to see him stand up for himself and his child. I don't say anything, frankly, because I don't have the energy.

"Fuck off, Tyler. You know I'm right. We're all starving to death—literally. We're wasting away out here." Cyrus snaps back.

"I told you to stop and go somewhere, Cyrus, and do it before you piss me off." Tyler's voice sounds more agitated.

"Well, hotshot, what are you going to do about it? Look at you. You've given all your food to that psycho bitch. You couldn't hurt me even if you tried."

"Cyrus, walk away." Tyler's voice is cold and stern,

reminding me of our father. *I've never seen this side of my brother before.*

"Make me," Cyrus says with a wicked grin, as if deliberately testing how far he can push Tyler before he snaps.

"This is your last chance. I said you need to walk away, Cyrus," Tyler demands. I hear crackling from outside and turn to see Xander outside, adding more pieces of broken furniture to the fire. The flames rise high into the sky, sending embers and sparks flying.

"And I told you to make me," Cyrus responds.

Just as Xander steps into the medical center, a blast of cold air rushes through once more, sending a chill down my spine, almost like an omen. A second later, Tyler sprints directly to Cyrus. Kelly screams. Julie walks in carrying a stack of books, completely unaware of the drama, and is knocked over by Tyler as he charges past. Xander steps back, holding the door open, and Tyler tackles Cyrus to the icy-cold deck.

Tyler slams his fist into Cyrus's jaw and crimson blood paints the white snow like an abstract painting. Cyrus fights back, knocking Tyler to the ground. Tyler's head bashes into the deck. Cyrus jumps up and slams his fist into Tyler's face, and with each hit, blood gushes.

"Cyrus, get off him!" I shout, running outside to pull him off. Tyler looks up at me, putting his hand up to stop me.

"I can handle this, Theo. Go back inside," he demands. Cyrus bashes him in the face again, and it takes everything in me not to intervene.

"Your whore and bastard child will die out here just like the rest of us, and there's nothing you can do."

I watch as Tyler grabs Cyrus's hand mid-punch, then shoves him off. Tyler quickly jumps to his feet, facing Cyrus, who is now lying flat on the ground. Tyler's face is bruised and battered, with one eye nearly swollen completely shut. Cyrus tries to stand but slips in the bloody snow and falls. As soon as Cyrus tries to stand again, I look over at Tyler and watch as he charges directly at Cyrus. I'm paralyzed in place, watching as my brother shoves Cyrus backward. I'm unable to speak as I watch Cyrus fall into the blazing fire. His clothes and hair ignite instantly. Cyrus lets out an agonized scream as the flames engulf him.

"Somebody help him!" screams Karissa. I reach out and cover her mouth.

"No, let him burn. If anyone wants to fight, it's going to be to the death. We don't have the strength or energy to put up with this," I say.

Karissa collapses to the floor next to me, tears streaming down her face as she watches Cyrus burn. I turn to see the rest of the group staring in horror. Still screaming, Cyrus tries to stand but his leg is caught on something and he can't move. Within seconds, he drops back into the fire pit and flails like a fish out of water, screaming. I stand, not offering or letting anyone else help.

Several minutes pass, the screams come to a stop, and his twitching body becomes still. The flames are so hot that his clothes are nearly singed off. His once-pale skin now has bright red patches and charred areas all over. His hair is completely gone, as are his eyelids, and his face burned with such intensity that his lips are also gone, revealing all his teeth. The raging flames flicker all around him as his body comes to rest, facing the medical center, and I swear his eyes are staring directly at me. Sara lets out a scream just feet away from me and drops to the floor. Then the room goes silent.

No one moves a muscle. A moment later, Tyler, with his swollen eyes, bloodied face, and split knuckles, opens the door.

The cold rush of wind blasts in once more. We all stand staring at him for another several minutes, processing what just happened. Tyler begins to breathe harder and faster, almost as if he is having a panic attack.

"I need some fresh air, I can't breathe," says Tyler as he opens the door behind him. This time, the cold air wafts inside with the scent of burning flesh. I glance around and realize no one is speaking because everyone's eyes are laser-focused on the burning body; no one is crying or yelling anymore. The smell is overwhelming, as is the sound of stomachs growling around the room.

"We can't do this," says Chris from behind Sara. I don't ask what he is talking about because I know exactly what he is referring to by looking at all the faces around me.

"Why can't we? Do you not smell that?" asks Xander, licking his lips repeatedly.

"Dude, it's cannibalism," Chris says.

Right before I speak up to agree with Chris, the smell of cooked meat hits me. My senses explode, and suddenly, my mouth begins to water.

"Chris," says Sara, drawing all our attention to her. She grabs his hands and lowers her head. "Chris, you are right. It is wrong. But this is now a matter of survival, and if we don't..." she pauses for a second. "If we don't eat, we will all starve to death. Think about it. You know it's true."

"Who are you? The Sara I know would never consider it," Chris murmurs, pulling his arm away with a look of disgust.

"Then don't do it, Chris," demands Xander. "It's more for the rest of us. I can smell it everywhere."

The door opens, and the smell wafts in once again. My mouth waters even more, almost to the point that I cannot think straight.

"I say we do it. I agree with you, Sara. This is a matter of survival. His body can feed us today and keep us alive for at least another week," I say, walking to the door.

"Don't do it," says Chris as he rushes past me and blocks the door. "You'll all regret this. I promise." I ignore him and push him out of the way. He falls to the ground, and I open the door, unable to control my hunger any longer.

47

MARY

Almost an hour later, Kelly and I stand back and watch as Theo and the rest of the group push their way outside and drop to the ground. I was only able to hold them back for that long by explaining that the body needs time to cook before they eat him.

Currently, they're all surrounding his body like vultures. Xander fills a bucket with snow and dumps it all over Cyrus' body and the firepit. The water sizzles and steam rises as the fire goes out. Unable to hold back any longer, Xander and Theo pull the body off the firepit. Each one grabs a body part and sinks their teeth in. Kelly and I both scrunch our noses in disgust as each person pulls their head back and the meat is ripped off the bone. Heat and steam rise in the cold air with each bite. The sound of chewing and swallowing fills the silence.

Kelly starts gagging and runs to the back of the room. When I look over, vomit spews from her mouth. Chris gives up and runs outside to join. As soon as he opens the door, the smell of meat wafts inside again. I fight the urge to run outside with him, even though my stomach is telling me to eat. I focus all my energy on Kelly and run over to her, dropping to the floor beside her. Gently, I rub her back, understanding her reaction. She looks up at me, then out the door to the group, and her eyes go wide. She turns back and vomits once more.

"Honey, they're starving; it's a survival mechanism at this point. Try not to look or think about it," I say, hoping it will help Kelly and convince me not to give in to the temptation.

I will sneak some of the saltines I stashed in the medical bag when no one is looking. I have to keep telling myself this, or else I know that I will give in.

"I know, but how... how could anyone eat another—" she doesn't finish her sentence. "Even if I were starving... I couldn't bring myself to do that."

"You say that now, but if you were truly starving, I think you just may consider it." I pause and let out a deep breath. Thankfully, the smell of Kelly's vomit is overpowering the smell of the meat, making it easier for me to focus on her. "You have to realize that you've been eating, so your body and mind aren't in the same state as the others. To them, they're imagining what they're eating is something else," I say, unsure whether what I'm telling her is actually true but hoping it is.

"Yeah, you might be right," Kelly responds. After a brief pause, we turn to face the group as they each stand. As the group moves toward us in the medical center, Cyrus's body comes into view, and I gasp. My mouth drops open. His body is nearly picked clean of all the tissue. They each walk back in and take a seat on the ground in the center of the room. Kelly and I both keep quiet. Kelly walks away and grabs a sheet to cover her vomit. Then she stops and stares at the ground, almost like she is lost in thought for a moment. When she turns back to me, she gives me a strange look. Almost like she wants to ask me a question.

"What? Why are you looking at me like that?" I ask, confused.

"Nothing," she says quickly, then looks away.

"It must be something. What's wrong?" I ask. My gut is telling me that Kelly is holding back as she nervously fidgets in place and looks away once more. "You can trust me, Kelly. Whatever it is, I'm sure it's okay."

"No, it's nothing, I promise." She goes silent for a moment, then smiles a fake smile. "I just wanted to say thank you for helping me," she says softly, and I can hear a slight tremor in her voice. But before I can say anything else, she walks away. I keep my eye on her as she walks over to Tyler. Before she gets to him, she glances back and catches me staring. She turns back around quickly, and when she reaches Tyler, I watch her frown and squeeze his arm, almost as if she's scared. And for the life of me, I have no idea why.

48

KELLY

"What's wrong, babe?" asks Tyler as I do my best to pull him closer to me, so that Mary can't see me.

"I can't believe you just did that," I say, disgusted. He doesn't say anything. I look over his shoulder and notice Mary is still staring at me.

"Don't look now, but I think Mary is hiding something," I whisper.

"What do you mean?" And just like I told him not to, he turns around to look at Mary.

"I told you not to look. Now she's going to know I'm talking about her." I glance over Tyler's shoulder one more time and feel a bit of relief when Mary is no longer staring at me and talking to Karissa instead.

"What makes you think that?" Tyler asks, pulling my attention back to him.

"I can't talk about it here," I say quickly. "Get Sara and meet me in the atrium." As soon as Tyler starts to protest, Theo clears his throat, pulling everyone's attention to him. He stands in the center of the room, wiping his mouth. It grosses me out, and I take shallow breaths to keep the bile in my stomach from rising again. He glances around at everyone, and I follow his line of sight, watching in disgust as the rest of the group all do the same —wiping their mouths and licking their lips.

"What just happened was necessary; we did it to survive. If you feel guilty, remind yourself of that. Repeat after me: we only did that to survive." I stay silent while the rest of the group repeats his words before he continues. "No one ever needs to know about this. No one needs to speak of what happened to Cyrus. I want to move his remains to the atrium and we'll tell everyone that he wanted his bones kept for his family. As for Brent and Megan, we'll tell the police it was suicide. Do I make myself clear?" No one responds. "I said, do I make myself clear?"

The entire room jerks at the force of his voice, and everyone responds at once, "Yes."

"Why don't we just toss the bones overboard and then no one will ever know?" I ask.

"Because I don't think that's a good idea. I mean, we ate him. Let's at least have a little respect and give the remains to his family," answers Theo.

"I'll move the bones," says Tyler. Everyone turns to look at us.

"I'll help," I say, thinking this would be the perfect way to get Tyler alone and tell him my thoughts about Mary.

"I'll help too," says Mary. I roll my eyes and let out a soft grunt of frustration.

"No, I think we have it under control; we'll be fine," Tyler responds.

"But I insist," Mary says, staring directly at me. I instantly feel heat rise in my cheeks. I don't want to look suspicious, so I smile. "I don't believe you should be doing all of that, under the circumstances, Kelly."

She's right, but can't anyone else help?

"Okay, then, Sara, I want you to come with us," I say, smiling. "Problem solved."

"Sure, I don't mind," Sara responds, although the look on her face suggests she'd prefer not to. I look at Mary, and she is still staring at me. It makes me uneasy.

We carefully take a sheet from inside and place Cyrus's remains on it. I stand there, refusing to touch any of it, grossed out just by being nearby. While we are outside in the cold, the door shuts. I look back and see everyone else, plastered up against the doors, watching our every move.

"What did you want to tell me?" asks Tyler, lifting the sheet with Cyrus's remains.

"I can't tell you here; I'll tell you in the atrium," I respond as quietly as I can, knowing that someone inside might hear what I say.

"What are y'all talking about?" asks Sara.

"I can't tell you out here, guys. I said I'll tell you in the atrium," I say, hoping they will drop it.

As the three of us walk through the medical center with Cyrus's remains, the entire room stares, and I feel like a spotlight is suddenly beaming on me. I glance over and see Mary still staring at me. I nervously look across the room and see both Julie and Theo, but they're both looking at Mary. That's weird. Maybe I'm not the only one who thinks something's up with her. Maybe I should ask him to come along as well. But before I have the chance to ask, Sara grabs my hand, and we walk out the door.

The atrium is cold and terrifyingly quiet as the creaking from the dilapidated chandelier echoes in the room. The body hanging swings back and forth with the rocking of the ship. The hair on my neck stands when I look up at it and remember the cracking sound his neck made when he fell. Maybe I should've picked somewhere else to talk. Oh well, we are here now.

"Okay, spill the beans — what's going on?" asks Sara, pulling my attention.

"Keep your voice down; I don't want anyone to hear us," I say.

"Okay, so what is it?" Tyler asks. I pause, glancing around to make sure no one else is around.

"I believe Mary is hiding something," I say quickly.

"Like what?" asks Tyler.

"I think Mary is the one taking the food."

"There is no way!" says Sara, laughing. "She's so sweet and always helps everyone. Why would she take food when it contradicts everything she stands for as a nurse?"

"Yeah, I have to agree with Sara on this one, babe. That doesn't sound like Mary," responds Tyler.

"I thought the same thing. That was until you all lost your minds and did... you know—" I say, glancing down at the bag with Cyrus's bones. "She didn't eat anything!"

"Yeah, and neither did you," says Tyler.

"That's my point! I didn't eat anything because I'm not starving—I've been drinking baby formula. When Mary and I were standing back while you all fed yourselves, she said you were only doing that out of survival instincts because your bodies were starving, and that I wouldn't understand because I'm not at that point."

"Okay? I still don't get it," says Sara.

"Oh my god! I understand what you're saying," Tyler replies.

"I'm still confused, so please explain it to me," says Sara.

"Mary wasn't eating because she wasn't starving. The reason I'm not starving is that I've been eating. If she's not starving, then what has she been eating?" Sara covers her mouth in surprise.

"Maybe, she couldn't bring herself to, you know... do that. She is a nurse, don't they take an oath or something? Wouldn't that go against that?" asks Sara.

"I think doctors take oaths, not nurses. Anyways, it's really suspicious, right?" I ask.

A sound comes from above, and as we all look up, all we see is strawberry blonde hair disappear.

"That was Julie," Tyler says.

"I think it was too," says Sara. "I wonder if she was listening all along to what we were saying."

"Only one way to find out," Tyler says, dropping the sheet with bones on the floor in the center of the atrium, and both he and Sara take off up the stairs, with me trailing along behind them. We all trudge up the stairs to the 6th floor. When we arrive, it's empty. I take a breath and hold on to the wall.

"Let's go back and see if she is still in the medical center," I say. "Go ahead; I'll catch up. I can't keep running like this."

"Sara, you go. I'll stay with Kelly," says Tyler.

"No. Let's stay together. We can walk slower," responds Sara. Together, we make our way there as fast as we can. I expect Sara to play it cool when we open the door and not cause a scene, then I remember who I'm talking about. As soon as the door opens, Sara brings everyone's attention to her.

"Julie, where are you?" shouts Sara.

"Is everything okay? Julie isn't in here; she was in the supply room then she left right after you all did," Chris responds. "What's going on?

"Nothing, where is Mary?"

"In the back with Theo," answers Chris.

"Oh my god, is it the baby?" Karissa gasps, looking at me.

Suddenly, everyone's attention turns to me, and I feel my cheeks grow hot. I hate when all the attention is on me.

"No, we're fine. I was just looking for them. I'm out of breath from running. Nothing to be alarmed about," I say with a fake smile, hoping it will make everyone stop looking at me.

From the back of the room, Theo and Mary walk out the door of the supplies closet, and all eyes turn to them.

"Why is everyone staring at me?" asks Theo.

"It's not you, it's Mary," says Sara.

"What did I do?" Mary asks, a look of confusion on her face.

"You know what you've done. Should I tell everyone?" says Sara.

"No, honestly, I have no idea what you're talking about," replies Mary. For a split second, I believe her. She genuinely looks like she has no idea. I close my eyes and take a deep breath. Deep down, I hope I didn't just create a world of trouble for someone who is innocent. But when I open my eyes, I see Mary looking me dead in the eye, and I can't stop myself.

"Mary has been eating our food; she's the reason we don't have any left. It had to be her stealing the food from all of us." The room gasps at the accusation.

"What? No, I didn't, I promise. Why would you say that?" asks Mary, her eyebrows raised high.

"Julie told me you opened those chips in the childcare room, and when she suggested sharing, you refused." Everyone in the room gasps again at Theo's comment.

"I did not; she is lying. She is the one who wouldn't share. I said we should bring it up to y'all. You've got to believe me," Mary says, pleading, on the verge of tears.

"Yeah, but it was you who found all the food on the 7th floor," shouts Xander. The room erupts in agreement.

"I promise it wasn't me," Mary begs again. "I can prove it."

"How?" responds Theo, and I perk up, interested in her answer.

"Where's Julie?" Mary asks. We all look around the room, and she is nowhere to be seen. Karissa walks up to Mary and snatches the bag of medical supplies from her.

"Are you hiding it here?" demands Karissa. The bag hits the floor, and all the supplies and medical equipment spill out. Everyone's eyes widen as a box of saltine crackers tumbles to the floor. I *knew* she was lying.

"I can explain that!" Mary shouts.

"Holy shit—you've betrayed us all," Theo practically growls, his face now flushed, his nose flaring with each breath. He steps away from Mary and stands beside Tyler and me. We all stand staring at Mary.

"I put that there when I found the supplies because I knew they'd go quickly. I thought it was a good idea to store some there with the supplies as an emergency backup. I haven't eaten any of them. Look, they aren't even open."

"Why should we believe you? All the evidence points to you," says Theo. He's right, this makes her look guilty as hell.

"Theo, I promise I can show you—there's so much you don't know. But we need to hurry."

"What do you mean by that?" asks Theo. What does she know that the rest of us don't? Suddenly, Mary drops to the ground, frantically scanning the medical supplies.

"Where are they?" she shouts, panicked.

"Where are what?" answers Theo.

"There was a glass vial in here with three syringes," responds Mary, and I notice her eyes fill with tears. "I promise I'm telling the truth. You have to believe me."

"What are you not telling us?" I cry out, confused about what is happening.

"I need to find the syringes. Please help me, and I'll tell you everything."

"No! First, tell us what you are hiding from us!" demands Theo, his voice sharp. I jump at the bass in his voice. No one notices the door creak open behind me; they're all too preoccupied with Mary. Suddenly, I feel Tyler squeezing my hand. I turn and scream as Julie jabs a needle into the side of both Theo's and Tyler's necks. As she pulls the needles out, I hear her whisper in Theo's ear, "I'm so sorry."

Neither Theo nor Tyler moves or speaks. Both are suddenly paralyzed in place. Tyler, weakened from not having eaten, drops to the floor first. I scream again. The entire room turns to see Julie standing there with one last syringe in her hand. No one moves.

"You bitch. See, everyone? I told you she was lying," shouts Mary, scrambling to her feet.

"If any of you come any closer, I'll kill you. I'm getting off this ship. I'm done. I won't stay trapped on this ship any longer," Julie says, pointing the needle towards everyone.

"What did you do to them?" I croak, slumping to the floor and cradling Tyler's head in my arms.

"I gave them something to make them sleep... forever. This last one is for you, Kelly. You psychotic bitch."

"No, don't," shouts Mary. "She doesn't deserve any of this. None of it is her fault."

"I would've spared her if she hadn't gotten knocked up by Tyler, but I can't have another Parrish to worry about," says Julie.

"What are you talking about?" I ask.

"Please, Julie, it doesn't have to be this way," Mary says, and from the corner of the room, I see Sara begin to move. Julie pulls a butcher knife from her pocket and throws it directly at her. The knife hits her with a sickening thud, embedding deep in

Sara's arm. I scream again, wanting this all to be over. I watch Sara drop to the floor, screaming in pain.

Theo tries to speak; his voice is raspy and barely audible. "Julie, why are you doing this?" Julie leans closer to Theo, meeting his gaze, and I see a tear roll down her face. It's almost as if she really doesn't want to be doing this.

"Your family has taken everything from me. I'm sorry it has to end this way," she says, bending down, kissing him one last time on the forehead. As she leans down, a folded sheet of paper falls to the floor.

Suddenly, Mary stands up and runs toward me. She shoves me out of the way just as Julie swings her hand. I watch as Mary hits the ground, nearly landing on her broken arm with a loud thud. Mary screams and rolls over, and my eyes go wide. Embedded in her stomach is the syringe— the one Julie had just said was for me. Mary just risked her life to save mine. But why? I was just blaming her for unspeakable acts. I feel rage building inside me, and I know I never cared for Julie. She just about killed me and my baby.

"Ahhhhh!" I scream at the top of my lungs. Sara screams, slowly pulling the blade out of her arm, then drops it on the floor. I heft myself up, grab the bloody knife, and charge directly for Julie. Within feet of her, Julie dodges me and sticks out her leg, causing me to trip. I fall on my stomach and feel white-hot pain sear through me. My hearing starts to fade, and the thought of Julie almost killing me and my baby overwhelms me. My eyes go dark for a moment, and I feel like I am passing out until I hear voices.

"Just leave, Julie. You did what you came here to do," Mary says weakly.

You did what you came here to do echoes in my mind. Was this planned all along? I open my eyes, and I'm surrounded by everyone. I look over and watch Julie run toward the door.

"Sophie? Kelly?" Sara shouts in my face.

"Yes, it's still me, Kelly," I answer, pushing Sara away, then scream again, feeling a sharp pain in my stomach.

"There is one more surprise I have for you all," Julie says, grinning as she opens the door. I see a large yellow bag with the words '*survival raft*' written in red on the side. She hoists it over her shoulder. "You'll all know what it is soon enough." Then she is gone. The door slams shut behind her. I continue to scream as the pain is excruciating.

"Karissa! Karissa!" shouts Mary.

"I'm here, Mary. I'm so sorry I didn't believe you," Karissa responds. I look over at Mary, who is lying on the ground.

"So am I. I'm so sorry I didn't believe you," I croak out.

"It's okay, sweetie," Mary whispers, before returning her attention to Karissa.

"The baby formula," Mary says weakly as whatever Julie just injected into her takes hold. "Get the baby formula!"

Karissa gives her a confused look, and I'm also confused about why she'd need baby formula at a time like this. Mary lifts her head just enough to point towards the closet. I glance over at the closet, then back to Mary again, and realize that she has nearly stopped breathing. She's trying to speak, but only gurgling noises come out. She is trying to tell me something, but I can't understand her.

"What do I do?" I ask in a panic. Mary coughs, trying to clear her throat.

"Antidote," Mary answers almost inaudibly. If I weren't right next to her, I would have missed it.

"I'll find it," Karissa yells, running to the supply closet. I slowly stand up, hoping the pain will subside, and when it does a little, I'm relieved. I follow Karissa over to the closet, and in the back, I see a small box. It's the only box in the closet, and it's labeled 'cleaning supplies.' Karissa quickly pulls it out and opens it, revealing two cans of baby formula. Karissa grabs a can and opens it—it's sealed with a silver aluminum lid that hasn't been broken. She grabs the other can, and something rattles inside. When Karissa lifts the lid, there's no powder; only a glass vial filled with clear liquid and three needles. I breathe a sigh of relief and follow Karissa back outside. Mary, Theo, and Tyler lie motionless on the floor. Every few seconds, they each take a shallow breath, showing they are still alive. Karissa kneels beside Mary, who isn't responding any longer, and I try to do the same, but with every move I make, a sharp pain runs through me, nearly crippling me. I do my best to breathe through the pain

"Is there anything I can do to help?" I ask, but Karrisa ignores me, focusing on Mary.

"Mary, wake up! What should I do?" shouts Karissa as she shakes Mary's body. Mary lies there, breathing short, shallow breaths like a fish slowly dying out of water.

"Mary, please wake up!" Karissa cries out. Mary doesn't move. Then Karissa raises her hand and slaps Mary across the face. The sudden jolt gets Mary's attention, and her eyes flutter.

"Oh, thank goodness! What should I do?" Karissa asks, holding the vial up for Mary to see.

Mary whispers, "Five... here," her hand slowly reaches up and points to the side of her neck.

"What does that mean?" asks Karissa.

I look at the syringe in Karissa's hand. "Maybe she means five units? On the syringe?"

Xander comes up behind me. "Yeah, that's probably what she meant."

Karissa inserts the needle into the vial and pulls back the plunger until it reaches five. After that, she flicks the syringe to remove any air bubbles and injects it into Mary's neck. Nothing happens. White-hot pain shoots through my stomach, and I let out a piercing scream. Karissa stares at me, and I grab my stomach.

"Help Theo and Tyler!" I scream through gritted teeth.

Without hesitation, Karissa rushes to them. I watch as she repeats the process and then injects into each of their necks. Moments later, Mary sucks in a deep breath beside me and sits up. A few seconds later, I watch Theo and Tyler do the same.

"Where is she?" shouts Theo, so angry that his veins bulge from his forehead.

"Oh fuck. It hurts so much!" *Please don't let me lose my baby.* I feel a wet sensation between my legs. I look down and see I'm soaked. I feel around, hoping it's not blood.

Tyler looks up at me, "What is happening?"

"Contractions...I think. When Julie pushed me—ughhhhh—it did something!" I scream.

"Oh my god," responds Tyler. I look over, and he looks like he is about to pass out again. The pain is nearly all-consuming now, and I drop to the ground, screaming. When I look up, I see Karissa, Sara, Chris, Theo, and Tyler standing around me. They all look like they are trying to help, but I don't know what to say. I can't think straight.

"I need you to breathe, Kelly," says Mary, grabbing my face.

"I am breathing," I snarl.

"I need someone to grab all the supplies," says Mary, still holding me.

"What are you doing?" asks Tyler, crouched down, squeezing my hand. I close my eyes and do my best to focus on breathing, in and out.

"I need all the supplies in case we need to move," says Mary.

"Wait, I can't move right now. What is she talking about?" I ask in a panic at the thought.

"Where are we going to need to move to?" asks Tyler. But before Mary has a chance to answer, there's a sudden explosion that shakes the entire ship.

"What the hell was that?" asks Tyler. I continue doing my best to control my breathing through the pain.

Tyler lets go of my hand, "Stay here, babe. I'll go find out what that sound was."

"No!" I demand through gritted teeth, grabbing his hand again. Let someone else look.

"I'll look. Tyler, you stay beside Kelly. She needs you," says Karissa.

49

THEO

My head is pounding, and my vision is still hazy. I can't believe my wife tried to kill me. What the fuck is happening? My body feels like it's on fire from whatever she injected me with. I have no idea what that explosion was, but I'm going to find out.

Opening the door to the patio, the cold air feels good on my skin. I feel like I'm burning from the inside out. I take a deep breath and steady myself, then walk to the railing and look up. The sky is filled with black smoke, and fiery orange flames flicker from the top deck as debris falls into the ocean. The strong wind blows the smoke across the sky. When I look down, I spot Julie on the 4th floor tossing a yellow bag into the water with a plastic bag tied to it. I'm shocked when I see her leap from the broken fourth-floor balcony, landing in the cold water below. After a few seconds, her head breaks the surface, and she grabs the yellow bag and tugs on a string. The bag bursts open into a round life raft. She struggles to climb aboard, eventually making it, and pulls the plastic bag inside. She quickly strips off her clothes and redresses with clothes that she pulls from the plastic bag. She then uses the remaining clothes to dry the raft, tossing the wet items overboard. When she's finished, she falls backward, staring up into the sky. She shifts her head and looks at the ship, never noticing me. My eyes remain fixed on her as she drifts further out to sea. My emotions running wild, I reach up to wipe the tears from my cheeks.

"Why did she try to kill me? None of this makes sense," I mutter, walking back inside the medical center, but no one responds. I glance over and see Kelly curled up on the floor in the center of the room, with Mary beside her. Xander walks past me to go outside.

"Did you see what the noise came from?" he asks. I don't answer, still stunned by everything that has happened.

"Oh my god," says Xander. I turn and face him as he looks over the edge of the ship, then out into the water. "What is she doing?" he asks. I walk closer to the window. As soon as I get near, I realize Xander is staring at Julie on the yellow raft drifting away from the ship, the dark cloud of smoke trailing in the same direction.

Overwhelmed by the chaos, I stumble backward from the window, needing to sit before I pass out. The sound of crumpling paper under my foot catches my attention, and when I look down, there is a piece of folded paper with my name written on it in Julies handwriting. My mind spins, and I can't think. I'm having a panic attack. Closing my eyes, I take a deep breath and hold it for a few seconds, then exhale. All the noise around, including Kelly's screams, fades into the background. Opening my eyes, I slump down onto the cold floor and unfold the paper.

Dear Theo,

You've been part of my life for a long time, and I know you better than I know myself, which is frightening. You're incredibly brave and always put others first. You see the good in everyone, even when no one else does. You've loved me through all our highs and lows. You were there when the doctors told me I could never have children. I've never told you the full truth about why I can't have children, or that even before that appointment, I already knew it wasn't possible for me. I'm sharing this now through this letter because I can't do it in person. There are many things about our past you don't know and were never meant to find out. Your father paid a lot of money to keep secrets hidden from you.

Over the years, I've watched you work so hard to impress your father and become the person he always wanted you to be, and that hurts me because you are nothing like him. His money and power turned him into a wicked man, and he has more secrets than you can imagine. I know you loved him more than anyone, so I don't want to ruin that for you, but I promised myself a long time ago I'd get revenge for what he did to me and my family. I also want you to understand that I don't blame you or Tyler for your father's actions. You were both just collateral damage, but it was the only way my plan could succeed.

We were never meant to meet, but fate had other plans. You don't remember the night that changed both our lives. Twenty years ago, your father took everything from me and my family. That night, you and your brother were involved in a car accident, and while you both walked away unharmed, your mother was killed by a drunk driver. At least, that's what the press reported. The police report that included the names of everyone involved that night went missing and the story the press reported never made sense to me.

So I did some digging, and found a high schooler who was at the scene. She was studying to be a journalist and wrote a detailed article about the accident. Most people never saw it because it was never published in the local paper, only on a high school website. The young girl was awarded a scholarship for her well-written article, entitled: "Fatal Accident of Mrs. Benjamin Parrish." To my surprise, I discovered that the scholarship she received was from the Parrish family. That seemed really strange to me, so I reached out to her, obsessed with finding out the truth. Every time I called, I used a fake name, but she refused to talk, saying it was years ago and she didn't remember. After months of failed attempts, I visited her in person. I told her I wanted the truth.

She was nervous at first, but I bribed her with the little money I had at the time, and she told me everything. She explained that the Parrish family offered her a full scholarship if she changed the article she originally wrote and removed it from the school paper. She didn't argue; she took the money and destroyed all the hard copies at the high school so no one could read them again. Then she rewrote the article, changing the information about the drunk driver. The Parrish family encouraged the fabricated account and awarded her a full scholarship to any college.

But when I met with her, she surprised me by showing me the original article she'd written, saying she knew someone would come looking for it someday. She'd kept the original version for insurance. I told her I'd return someday with a lot more money after I got my hands on it.

My mother was the so-called 'drunk driver' who killed your mother, but she'd never drank a day in her life. Your father paid everyone off to cover up the fact that your beloved mother was an alcoholic who got behind the wheel with her twin ten-year-old boys in the car. The town would have gone ballistic, and your father couldn't risk destroying his family's reputation.

I knew the police would never act on what I found, so I made a promise to seek revenge on the Parrish family, no matter the cost.

The woman who wrote the article managed to get an interview with a paramedic at the accident that night. I tracked down that paramedic and learned she'd become a nurse. When I found her, I made her an offer: help me take down the Parrish family for what they'd done, and I'd give her ten percent of their wealth—more money than she'd ever need. She asked me how I'd do that. I told her I was going to marry one of the Parrish brothers, and once I'd settled in and joined the family, I'd kill them both, thus making me the sole beneficiary of the Parrish inheritance. She thought I was crazy—and maybe I am. But with the amount I offered, she couldn't refuse.

Thanks to me, she managed to get an invitation on the ship. Mary, our beloved nurse, was the one who set off the bomb I planted. It was meant to kill you on the bridge the first night we set sail because I couldn't bring myself to do it. All of this had been planned for so long, but I actually fell in love with you, which was never part of the plan.

I asked Mary to kill you, and I'd take care of Tyler. If that storm hadn't happened and we had stayed on course, you would have died, I would have poisoned Tyler and thrown him overboard, and I would have inherited everything. But that damn storm ruined all my plans. We've been trapped on this ship for months, and now I have another blood relative to worry about—Kelly's baby. I can't let that little shit take what's mine. I wish things could have been different, but you can't always get what you want.

Julie

50

THEO

I drop the letter, my body weak from the overwhelming emotions flooding my mind. The sound of Kelly screaming and Tyler shouting echoes in my ears. I stand up and walk over to Mary, shoving her to the floor. I'm blinded by rage. I could snap her neck right now.

"What the hell, Theo? What are you doing?" Tyler shouts.

"This is all her fault. *All* of this is her fault." I can barely keep it together, my heart pounds in my chest. "Mary, I should kill you right now, you lying bitch."

"Theo, what's happening? What did she do?" Tyler asks. I look down and see him holding Kelly.

"Get the fuck off me. Where is my husband?"

Oh fuck, not this again.

"Great, now we have to deal with Sophie in the middle of this," Sara mutters, throwing her hands up as she rushes over to comfort Sophie. My attention snaps back to Mary. I walk closer and closer as she scoots backward, looking up at me.

"What did I do?" begs Mary, and I can tell she is terrified. It takes everything in me not to reach down and break her arm again.

"Tell the group what you did. Tell them why we're all trapped in this nightmare because of you."

"What is he talking about, Mary?" Tyler asks.

"I can explain; it wasn't supposed to happen this way," Mary pleads.

"What wasn't supposed to happen this way?" Xander asks.

"Tell them what you did." It's taking everything in me not to choke her right now.

"Wait, Mary too?" shouts Tyler, sounding confused.

"Yes, tell them everything you did before I throw you overboard. No one will stop me once they learn the truth. Tell them!"

"Okay, I promise it wasn't meant to happen like this. The explosion that took out the ship wasn't an accident." Everyone gasps, and Kel– or, I guess, Sophie now– lets out a scream.

"Guys, I think the baby is coming," shouts Sara. But now everyone else is too focused on Mary to look away.

"The explosion was meant to kill Theo and disrupt communications," Mary says, causing everyone to gasp again. "But then the storm came out of nowhere, pushing us off course. After that, we got lost. All of this was Julie's idea; she'd been planning it for years."

"What do you mean she's been planning this for years?" Tyler asks.

"She reached out to me years ago with an offer to help take down the Parrish family," says Mary.

"How? When Julie met Theo, she had nothing," says Tyler, then he turns to look at me.

"She said she'd pay me once she killed you and Theo and inherited all the money," Mary responds.

"She was going to kill me, too?" asks Tyler.

"Yes! It's all right here in the letter," I answer, shoving the papers at Tyler.

"Your mother killed Julie's mother in that car accident twenty years ago, and your father orchestrated a cover-up that framed Julie's mother as the drunk driver in the accident," says Mary. "I begged her for months not to go through with it, but she was hell-bent on getting revenge."

We all stand there in stunned silence until a guttural groan rips through the air.

"Guys... I really think the baby is coming," says Sara.

"Grab the medical bag!" Mary calls out.

"Shut the fuck up," Tyler snaps right back at her.

"Look, I get that you're all pissed at me, but let me help her. Please," says Mary. Everyone else springs into motion, but I can't move. The world suddenly feels like it's moving in slow motion.

"I think you've already done enough," Sara snaps. I turn to Sophie and watch as she grunts, clenching her teeth and grasping at Sara's arms.

"Where is my damn husband?" asks Sophie. Suddenly, she stops screaming, and her eyes roll back.

"What is happening?" asks Tyler.

"Oh my god, Kelly, wake up," screams Sara. "Mary, what's happening?"

"She fainted," says Mary. From the look on her face, I'm not sure if she actually knows why, though.

"Ughhh—oh my god, is the baby coming?" groans Kelly or Sophie, I'm not sure at this moment.

"Kelly?" asks Sara.

"Yes—uggghh—it's me," Kelly screams. I watch Tyler run over and kiss her forehead.

"Check to see if you can see the baby crowning," Mary says from across the room.

"Come over here, Mary, please," Kelly begs.

"No," demands Tyler.

"Why?" Kelly groans.

"Because sh–"

"I don't care. Mary, please—ahhh—get over here!"

Mary stands to go over, and I glare at her with a death stare.

"This isn't over," I say as Mary rushes past me to Kelly.

"Tyler, help me get her pants off," says Mary. Tyler hesitates at first, then drops down and does what she says. "She isn't fully dilated, so the baby isn't ready to come out yet, I can't do this with one hand. Sara, come help me."

We watch Kelly labor for what feels like an eternity, doing everything possible to keep her comfortable. By nightfall, the fire on the top deck has grown so intense that the patio glows orange against the night sky, and the back wall of the medical center is beginning to collapse.

"We have to get out of here. We have to get off this ship," I say.

"Where are we going to go? All the lifeboats are either taken or damaged beyond repair. We've tried everything. We have nothing," shouts Xander.

"Wait!" screams Mary. "The raft Julie had—I've seen it before. I know where she got it. I think there might be another one."

"Where?" asks Xander as the ceiling in the back of the medical center slowly collapses, letting in fire and smoke.

"The 1st floor in the crew area, where Captain Whitaker's body was—bring a flashlight because it's completely dark down there," shouts Mary.

"I know exactly where that is. I'll go with you. Come on, Xander, let's go," I say.

As we rush toward the door and head for the stairs, the sound of Mary's voice echoes in the hallway.

"We'll meet you at the fourth-floor balcony at the broken railing. Please hurry."

51

TYLER

"Grab as many clothes as you can, all the sheets and blankets, too, and put them in a trash bag. Tie them securely so they don't get wet. We are going to need all the coverage we can get. Also, grab the baby formula and stuff as many bottles of water as you can into the medical bag," Mary shouts to Karissa.

"What are we doing?" asks Sara. I look over and see Chris standing next to her, pale and looking like he is about to puke.

"We're getting off this ship," says Mary.

"What?" Kelly screams. "I can't get off this ship. I can't even walk."

"I'll help you, babe," I say. Kelly squeezes my hand so tight that I feel it going numb. I smile, trying my best not to pull away.

"We have to get out of here now. Karissa, did you get everything?" Mary asks.

"I think I did," answers Karissa, holding up two large, clear garbage bags completely filled.

"Chris, help her carry those to the 4th floor. Double or triple bag them so they don't rip," Mary says, then points to Sara. "Sara, help Tyler carry Kelly to the 4th floor. I'll bring the medical bag." I do what Mary says and pick up Kelly. She groans again, and I wish there were more I could do to help. Luckily, we are going down the stairs, not up.

When we reach the 4th floor, it's dark and cold. Stepping outside, I set Kelly down. The wind is howling, and the broken lifeboat sways. Out on the open ocean, the bright orange glow reflects off the rough seas. The fire on the top deck burns more fiercely than ever, completely engulfing the ship. In every direction, there is nothing visible for miles except black smoke.

"Where are Xander and Theo?" asks Chris, standing to the side with Karissa and holding one of the bags.

"They'll come; give them time," Mary screams as the wind whips around us.

"Oh my god, this baby is gonna–" Kelly gasps and grips my shoulder tight, her fingers digging into my skin. I bite back a wince.

She pants for a few seconds, then tries again. "This hurts so much," she groans as she breathes rapidly. She squats down as if she is trying to bear down.

"Babe, what are you doing?"

"No, Kelly, don't push yet," screams Mary.

Suddenly, Theo and Xander come running out the door, out of breath, and drop the raft on the deck in front of us.

"The wind is blowing in the opposite direction," says Theo, staring up at the smoke-filled sky.

"What do we do now?" Xander asks, and we all gaze at the open ocean before looking down the side of the deck, which is easily forty feet below.

"Jump. We have to jump," says Mary.

"What? How? What about the baby?" I ask.

"We need to pray that everything goes well. We can't stay on the ship; it might burn through, and we'll die from smoke inhalation."

"Can't we just wait it out? Surely rescue will see this fire," shouts Xander, his voice cracking in fear, looking down at the water. As he speaks, burning debris from the top deck cracks off and plummets into the water. Bits of burning material catch the wind and fall onto the 4th floor, just feet away.

"We've been out here for months, Xander. No one's coming, not now," says Theo, his shoulders drop, and he turns to look at me. *Was rescue ever coming?* Kelly squeezes my hand, pulling me from my thoughts.

"I don't want to chance it," shouts Chris.

"Well, I don't want my baby to die," Kelly cries.

"We'll jump together, I've got you. But we've got to go now," I say.

The fire intensifies and burns brighter as the wind blows, scattering more debris that lights up the night sky.

"Okay, we will do it together. I love you so much," Kelly whispers in my ear.

"I love you more, babe," I say, kissing her. She screams in pain, clutching her stomach, and that's when I know we have to do it sooner rather than later.

"Well, if we're doing it, let's do it before this baby comes out," shouts Kelly.

Xander is the first to jump, followed by Karissa. Together, they inflate the survival raft. Once they climb in, shivering from the cold water, Chris throws the bags of clothing and blankets. Xander and Karissa quickly pull the bags into the raft. Sara, Chris and Mary are next, jumping together after counting down. Kelly and I are next. Before we jump, Theo stops us.

"I'm sorry we are in this situation. In case I don't make it–"

"No, Theo. We are going to make it."

"Tyler, I need to say this... please," he asks, tears forming in his eyes. "In case I don't make it, I want you to know that I love you more than anything in this world. You are going to be the best father a kid could ever ask for, much better than ours."

I stand stunned at his words. Kelly groans, pulling my attention back.

"I love you too, but we can go over how good a dad I'll be later when we are home," I say, wiping a tear from my cheek. "Now get your ass over here."

We face each other, and I look at my brother and take a deep breath. He looks at Kelly, then over at me, then nods. Kelly lets out another groan and pants, tears in her eyes, gripping my shoulders, her body trembling.

"I can do this," she says, her voice wavering.

"Yeah, you can," I say, as I wrap my arms around her. Theo grabs her from behind, and we brace her between us to help cushion the impact of the water.

Then we jump.

As we hit the water, both the impact and the utter cold of the freezing ocean almost takes my breath away. As we break through the surface, I thrash around looking for my brother and Kelly, and feel a little bit of relief as Kelly surfaces near me.

I look toward the raft, and everyone is shouting with their hands out to grab us. Quickly, we swim to the raft. Xander and Sara pull Kelly in first, laying her on her back. Immediately, they throw dry clothing over us to dry us and warm us up.

"The baby is coming," Sara screams as the raft fills with a mixture of blood and water.

"We need to deliver this baby now," says Mary as she yanks supplies out of her bag.

"Where's Theo?" asks Karissa. My heart sinks as I realize he didn't crawl into the raft with us. I turn quickly, scanning the water, but he's nowhere to be found. Several seconds pass with no sign of him anywhere. I can't tell if it's the cold wind or the state of panic I'm in causing my ears to ring. *Where is he?* A second later, Theo's body surfaces only a few feet away, face down.

"Theo!" I shout, diving into the water. Theo is unconscious, his skin pale, his lips tinged blue. I swim back to the raft with Theo. Xander and Karissa grab his limp body and pull him up. I follow. Xander immediately begins chest compressions, but it's nearly impossible with the raft shifting like a waterbed beneath him.

"The baby, I see it!" shouts Sara. My heart is pounding so hard that I feel it in my ears. My teeth are chattering so hard that I'm afraid I'm going to chip a tooth. I grab a piece of tattered clothing to pat myself dry and try to warm up. I glance over, and Sara is pulling Kelly's pants down, and all I can see is the bloodied, purple crown of dark hair. Feeling dizzy and on the verge of fainting, I take a deep breath and slowly crawl over to help.

"What can I do?" I ask. Xander continues doing compressions on my brother, and I can hear him panting loudly. He's gradually becoming more exhausted and out of breath as the rough ocean keeps knocking him over.

"Push, Kelly, push!" shouts Mary. Sara is between Kelly's legs, doing everything Mary tells her to do.

Kelly screams and I watch her head fall backward. A moment later, the sound of a crying baby fills my ears.

"It's a girl!" shouts Sara.

Mary grabs the medical bag and flings it open. "Where are the scissors?" No one responds, and Kelly slowly lifts her exhausted head, smiling at me.

"Can I hold her?" Kelly asks.

"Of course," says Mary. "Tyler, I need you to bite the umbilical cord. We don't have scissors." I stare blankly at Mary, not sure if I heard her correctly.

"I'm sorry, what did you say?" I ask breathlessly, because

there is no way she just asked me to bite the umbilical cord. The thought alone makes me gag.

"Bite the cord Tyler, you have to do it."

Slowly, I move closer to Kelly. I do my best to avoid disrupting Xander, who is still doing compressions. I notice he is almost falling over after each one. I take another deep breath and grab the umbilical cord, lifting it to my mouth, about to cut it with my teeth.

"Here, I found the scissors!" screams Sara. Oh, thank God! I quickly grab them and cut the cord. Sara hands Kelly the baby. Karissa rips open a bag of sheets and wraps the baby, then throws another one over them both.

On the other side of the raft, Xander lets Chris take over and collapses back into the raft. Just that second, Theo turns over, sputters, then vomits violently. Xander slumps down with a weak laugh and Chris sits back on his heels. I look over at Theo and let out a breath of relief, then I grab Kelly and kiss my newborn baby girl on the head.

Kelly smiles up at me exhausted, and I kiss her again. I feel tears forming in my eyes as I look back down at our baby girl. "What should we name her?" I whisper.

"Hope."

EPILOGUE

MARCH 2013

With a loud blast, a ship's horn lets out a piercing *bwaah... bwaah,* and the low rumble in the water catches my attention for a brief second. This isn't the first time my mind has tricked me into thinking I was being rescued, so I sink back into myself. My eyes are almost entirely closed, and my skin feels like it's peeling off from the cold windburn and the dried saltwater. My body is so weak that I don't have the strength to lift my head and look around. Moments later, I feel icy cold water splashing on my legs, and the sound of a man yelling in my ear jolts me awake.

"Hello, can you hear me?"

I struggle to open my eyes; the sunlight is blinding. After my eyes adjust, I see a man in a jumpsuit and safety gear leaning over me. I'm shocked by the sight, and my dry throat burns as I gasp. My ears are ringing, and I feel like I'm underwater for a moment.

The man shouts again, "Can you hear me?"

Realizing this isn't a dream, I croak out, "Are you real?"

The man touches my arm and answers, "Sure am. My name is Brandon. I'm with the US Coast Guard. You're safe now. We've got a helicopter on the way to take you to a hospital."

I'm still in shock, so I just stare at him blankly. The man grabs my shoulder, and I flinch. He speaks softly, "Everything's going to be okay. Can you tell me your name?"

I'm still too stunned to speak.

He tries again, "Do you have someone we can call?"

This time, I managed to get the words out.

"No, I don't have anyone left. They're all dead."

ONE WEEK LATER

Beep. Beep. Beep. The sound of a monitor wakes me. My entire body hurts, and my eyes sting. Looking down, I see an IV sticking out of my arm — I'm in a hospital room.

"Hi there," a voice calls out. I look up to see a man in a police uniform standing beside me. "Can you hear me?" His voice booms in my ears. My entire body is aching.

"Yes." My voice is hoarse, and my throat is killing me. "Can I have some water?" I ask softly.

"Back away from my patient, officer," commands a tall woman in a white lab coat, approaching me quickly. "Welcome back. You've been through quite a lot, haven't you?" I stare blankly for a second and all I can think about is how badly I need a glass of water.

"Water, please?" I beg, looking around for anything with a date on it.

The woman brings a cup with a straw up to my mouth and cautions me to take small sips. When she pulls away, I ask what day it is.

"My name is Dr. Jackson. I've been taking care of you. It's

March 17, 2014. You were found freezing in a raft in the ocean."

"That's why we're here, ma'am," the officer interjects. "Could you answer a few questions for us?" He has a belly that seems to be in a fight with his buttons to keep his shirt closed.

"Can this wait, officer? She just woke up from an extremely traumatic event and I need to run some tests," says Dr. Jackson.

"I'm sorry, but it can't. This is an ongoing investigation involving a large number of deaths and missing persons. We need to ask your patient some questions," says the officer.

Dr. Jackson huffs, rolling her eyes. "Fine, but please make it quick."

"I'm Officer Phillips, and this is my partner, Officer Cade. We have a few questions for you, if you're up to it?"

"Okay," I say, taking a deep breath.

"Can you start off by telling us your name?" asks Officer Phillips.

"Julie. It's Julie Parrish," I say.

"Parrish? As in the cruise line?" asks the officer, standing up straight and adjusting his belt.

"Yes," I say.

I watch as Officer Phillips glances at Officer Cade and lifts his eyebrows, then Officer Cade inclines his head at him.

"Get the chief on the line immediately. We can't let this get out. The press is already flocking around the hospital trying to get interviews. All hell's gonna break loose if it gets out that the only survivor is a member of the Parrish family."

"Understood," replies Officer Cade.

Officer Phillips pulls out a pad and pen. "All right, Mrs. Parrish, start from the very beginning, tell me everything you remember."

I feel my heart pounding as the sound of the beeping

monitor starts to ring in my head. "Okay... but I want a lawyer first."

ONE YEAR LATER

Sitting alone on my couch, I sip red wine and stare out the window of my Malibu beach house. Even though I'll never step foot in the ocean again, I can't help but love the view. The sunsets are to die for. The best part about living in Malibu is that I'll never have to deal with freezing temperatures again. I swear to God, I never want to see snow again in my life.

I feel the smooth texture of my white leather sofa, imported from Italy. I can't help but smile, thinking about how I got away with everything, how it all worked out even though my plan was essentially ruined. I'm glad Mary never made it off that ship alive; she didn't deserve any of my money. There is justice in this world after all.

I settle back into the couch and grab the remote, flicking my TV on. The 55-inch flat screen comes to life, and to my surprise, the movie *The Last Song* is playing. Instantly, my mind goes back to that letter I wrote to Theo. Then I think about all the police reports and news reporters I had to lie to over and over again.

I told the police everything — everything except the part where I killed my husband and his brother, and that I left the rest of them to die in a fire I'd started. I told them everything they needed to paint me as the sad widow who lost everything in what was supposed to be our family's happiest moment— the first sailing of the family's cruise line. For six months straight, I had to deal with all the legal bullshit of the cruise line paying out the victims. The disaster nearly bankrupted the insurance company and the Parrish Foundation, but thankfully, the family trust couldn't be touched because it was established well before the cruise line was ever created. I got everything I wanted, and I couldn't be happier.

Not wanting to dwell on the past any longer, I flip the channel. When I see the headline on the screen, my body goes weak, and the glass of red wine slips out of my hand. I look down and watch it land on the white sofa, staining it crimson before shattering against the tile floor. I look back at the TV. Suddenly, I'm lightheaded, and the room is spinning. Even though I'm staring at the television, I can no longer hear it. The only thing I can do is stand and stare at the words written across the bottom of the screen.

Breaking news— survivors from the Parrish cruise line disaster, including a toddler, have been found on an island.

Banging on the front door echoes in my ears, pulling me out of my trance.

"This is the police. Open up!"

I stand there, too shocked to move or say anything.

How the fuck did this happen? How in the hell did they get off that fucking ship? A loud bang sends debris flying all around me as police break down my door. I assume guns are raised, all pointing straight at my back, but I am still too shocked to react.

"Julie Parrish, you are under arrest for—" The voices fade, and I'm shoved to the floor. My hands are pulled behind my back, and cold cuffs are tightened around my wrists. The officers quickly pull me to my feet and escort me toward the door.

As soon as we step outside, my knees go weak when I see who is staring at me.

"How are you still alive?"

THEO

"Does it matter? I'm here now, thanks to an antidote and a deserted island. You failed, bitch."

"It's your word against mine. No one will believe you," Julie cries. I can see right through the dramatics.

"They will when I show them the proof that you planned it all," I say, grinning.

"You don't have any proof," she growls.

I laugh, pulling an envelope out of my back pocket. Carefully, I open it and slide the folded papers out. Julie turns a shade of gray and the police officer shoves her past me toward the police cruiser.

"I want to see my lawyer," she yells.

"You mean your public defender? You have no money," I say. A news helicopter hovers above us, and the driveway is lined with countless reporters. As the officers walk her to the patrol car, reporters point their cameras and hold microphones inches from her face, asking the same questions.

"Why did you do it?"

"Did you really believe you could get away with it?"

Then she is shoved headfirst into the squad car. Before the officer slams the door closed, he bends down and looks her right in the eyes. "You really messed up, sweetie. They're going to put you away for a long time," he says, then he slams the door, gets into the driver's seat, and drives away. Another officer walks up behind me and taps me on the shoulder. I turn around, glad that none of the reporters have figured out who I am yet.

"Can I take you back to the hospital with the others?" he asks. I stand there silently for a moment. I still can't believe that just happened. I look down at the letter in my hand, then fold it up and put it away.

"Please, but before we do, I saw a diner on the way here a few miles back. Do you mind if we stop there? I could really go for a burger."

When we arrive at the diner, the thought of a freshly cooked burger has my mouth watering. I slide into the booth, and a young waitress comes over to take my order. While I sit alone waiting for my food, I hear a voice from a few feet away say, "Can you believe what happened on that ship? This is why I'll never go on a cruise! I can't imagine what those people went through. How in the world does something like that even happen in this day and age?" When I look to see who said it, I realize it's an older redhead talking to the younger waitress who took my order. I pull my hat down just a bit more and slide on my sunglasses, hoping that neither of them will recognize me.

If she thinks the stories of what happened on the ship were crazy, wait until she finds out what happened on the island.

The end...
For now.

ACKNOWLEDGMENTS

For as far back as I can remember, I have had a vivid imagination. Even with how my mind always wandered, I never expected to write a book. But one day, a small idea played in my mind like a movie, and I knew I had to get it down on paper. After I wrote the prologue, I shared it with a few people, and they encouraged me to keep going. I didn't have any characters or even a full plot in mind, but I knew exactly how I wanted the story to end. I began writing the full story in June of 2025, and just four months later, my first draft was complete. Terrified but excited, I sent it off to an editor and never looked back.

I've always thought that writing a novel was a journey you take on your own. For me, that simply wasn't true. There are so many people who have supported, encouraged, and believed in me, and this story, from the very beginning. I am, and will forever be, grateful to those who supported me on this wild journey.

To my husband Tommy, thank you for your endless love, patience, and encouragement. Thank you for always pushing me to be the best that I can be, thank you for telling me to keep going every time, even when I convinced myself that people wouldn't like my work. You believed in me every step of the way, even when I didn't believe in myself. I love you, always and forever.

To my son Brycen, thank you for reminding me every day why dreams are worth chasing. I hope this book shows you that with enough love and determination, anything is possible.

To my mom, thank you for always believing in me and listening to me read every chapter to you as it was written. A core memory was created, and I will cherish those moments forever.

To my friends, thank you for cheering me on, celebrating every milestone, and reminding me to keep going when the finish line felt so far away.

To my editors, Anette King and Heather Caryn, thank you for helping shape this story into the best version of itself. You both made this book stronger than I could have imagined.

To everyone who believed in this story before it was finished, shared a post, preordered a copy, recommended it to a friend, or simply took a chance on a first-time author, I thank you from the bottom of my heart. Your support means more than words can express.

And finally, to you, the reader: thank you for picking up IF THEY KNEW. I hope this story kept you turning pages late into the night. I hope the twists and turns throughout the book surprised you when you least expected it, and I hope this story stays with you long after you've reached the final page.

Until the next chapter...
A.J. Wooding

ABOUT THE AUTHOR

Austin was born and raised in Plant City, Florida, a small town known as the Winter Strawberry Capital of the World. After meeting his now-husband, he moved to Orlando, and their journey eventually led them to Delaware. Austin began his professional life as a dental assistant, but after years of working in dentistry, he decided to chase adventure and became a flight attendant.

When he's not jet-setting across the globe, Austin is at home with his family: his husband, their child, and two spirited pit bull terriers. An avid reader and storyteller, Austin enjoys writing dark, suspenseful thrillers that keep you on the edge of your seat. Off the page, Austin's happiest moments are spent with his family and taking his son to Disney World or planning their next Disney cruise—a family tradition filled with laughter, magic, and always an unforgettable adventure.

Follow me on social media, visit Ajwooding.com

www.ingramcontent.com/pod-product-compliance
Lightning Source LLC
LaVergne TN
LVHW102153101026
849957LV00026B/175